A BOOK OF BEES

"Delightful. . . . Has an elegance that draws its strength from an understated prose style that rolls along without so much as a single awkward moment."

The Philadelphia Inquirer

"Sue Hubbell's writing . . . makes beekeeping—and the delights of a solitary woman living in rhythm with the land and its tiny, productive creatures—so appealing."

New Directions for Women

A BOOK OF BEES

Also by Sue Hubbell

FAR FLUNG HUBBELL

BROADSIDES FROM THE OTHER ORDERS

A COUNTRY YEAR

ON THIS HILLTOP

A BOOK OF BEES

. . . and How to Keep Them

Sue Hubbell

Drawings by Sam Potthoff

A Mariner Book

HOUGHTON MIFFLIN COMPANY

BOSTON • NEW YORK

Library of Congress Cataloging-in-Publication Data is available.
ISBN 0-395-88324-5

Printed in the United States of America

QUM 10 9 8 7 6 5 4 3 2 1

Portions of this work have appeared in *The New Yorker.*

Grateful acknowledgment is made to the following for permission to reprint previously published material: *The New Yorker*: The poem entitled "The Song of the Queen Bee" by E.B. White from *Poems & Sketches of E.B. White* (Harper & Row). Copyright 1945, © 1973 E.B. White. Originally in *The New Yorker.* Time, Inc.: An article entitled "In Missouri: The Cicada's Song" by Sue Hubbell, which appeared in *Time,* July 15, 1985. Copyright © Time, Inc. All rights reserved. Reprinted by permission from *Time.*

FOR THE SWEET BEE,
Apis mellifera

Beekeeping is a business that requires
the greatest amount of attention to
small details. . . . The good beekeeper
is generally more or less cranky.

—C. P. DADANT

Acknowledgments

The idea for this book came from Pam Strickler, an editor at Ballantine Books, and I want to thank her for suggesting it. But, in addition, it would not have been written had not Liz Darhansoff, my agent, known when to speak and who should not speak of it. Jean-Isabel McNutt at Random House has contributed to the book; she has enormous zest for locating obscure source material and a good sense of what my writing style should be.

A number of people have read the manuscript and made fine suggestions: Asher Treat, Liddy and Brian Hubbell and Arne Sieverts were all helpful.

The Hockman brothers, Dwain and David, helped with the descriptions of tools. Hugs, guys.

Contents

I

THE BEEKEEPER'S AUTUMN

The Beekeeper's Autumn

For a long, long time—for nearly forty years—I never had any bees. I can't think why. Everyone should have two or three hives of bees. Bees are easier to keep than a dog or a cat. They are more interesting than gerbils. They can be kept anywhere. A well-known New York City publisher keeps bees on the terrace of his Upper East Side penthouse, where they happily work the flowers in Central Park.

I have had bees now for fifteen years, and my life is the better for it. I operate a beekeeping and honey-producing farm in the Ozark Mountains of southern Missouri. I keep three hundred hives of bees, separated into groups of ten or twelve, in what are called outyards—land that I rent from other farmers at the cost of a gallon of honey a year, rent I pay to the farmers for the privilege of putting the bees there. The farmers and their families like the honey, but they like having the bees on their land even better. The clover in their pastures is more luxuriant because the bees are there to pollinate it, and the vegetables in their gardens and the fruit on their trees benefit from the bees, too. My best and most productive beeyards, however, are those near towns, because townspeople plant flowers and water both their flowers and clover-scattered lawns, providing the bees with a constant supply of fresh blossoms to secrete nectar which they turn into honey.

Every once in a while I read in the beekeeping magazines about someone who has had complaints about his bees. I am always astonished, because around here everyone has a fine friendly feeling toward them. My own beekeeping operation is a matter of minor local pride, and is the focus of interest and curiosity. People come out to my farm and ask if they may tour the "honey factory." I am asked to speak to local civic groups and high school biology classes. The bees themselves are regarded with a certain amount of affection and good humor.

The town in which I live is very small. All the other farmers raise pigs and cattle, and making a living from bees does give them something to talk about down at the café other than fescue foot and the price of pork bellies. Cows and pigs are large animals, and the farmers keep track of them by putting a numbered ear tag on each beast's ear. It tickles their fancies that someone can make a living with a bunch of wild bugs who can't be penned and marked, but who fly everywhere, unruly but helpful, pollinating plants and making honey. They enjoy telling jokes on me, I know.

Nelson is the town wit. Like any Ozark storyteller, he piles outrage on top of outrage without even the smallest trace of a smile. It was Nelson who, straightfaced, spread it around town that when a swarm of bees gathered on my mailbox and stayed there for several days, it was because I hadn't put enough postage stamps under their wings. Nelson said it was a well-known fact that with proper postage a bee could travel anywhere in the continental U.S. of A. " 'Course if they was to go abroad, the rate is a mite higher. Maybe they was. Seems like a smart lady like Sue ought to know how much postage to put on a bee."

I was sitting in the café one day with Nelson and some of the other good old boys, when Nelson, deadpan, said, "Say, one of your bees was over a-bothering my peach tree this morning."

"How'd you know she was mine, Nelson?" I asked, looking him straight in the eye. I was determined to put up a fight this time.

Nelson hadn't expected my question.

"Why, they're all yours, aren't they? I thought you owned every blessed bee around here, the way you're always talking 'em up at the Chamber of Commerce meetings."

"No, that isn't true. There are wild ones in trees all over, and then Henry has some, and so does Billy, right here in town. I'll tell you what you've got to do before you carry on so about *my* bees bothering your peach tree. What you've got to do, Nelson, is go over to that tree and check the ear tag on the bee. The first thing I do after a new bee is born is to put an ear tag on her ear. You check the tag and you tell me her number, and *then* I'll let you know if I'm going to accept responsibility for that bee."

Nelson threw back his head and laughed.

A report of our exchange went around the café. This was several years ago, but even now every once in a while someone will stop me in town and say, "Hey, Bee Lady, I saw Number three fifty-seven on a clover blossom in my lawn today."

"That so? How's she looking?"

"Jest fine. Better in fact than the last time I saw her, when I thought she looked a mite peaked."

"Good to hear she's improved."

The end of one honey season is the start of the next, and autumn is a good time to begin with bees. It is when many people buy a few hives from an established beekeeper and move them from his place to theirs, because at that time a beekeeper is often willing to sell some of his hives at a lower price than he would in springtime. He has already taken his honey crop, and there is always a certain amount of risk in carrying a hive through the winter, a risk that is transferred to the buyer.

Summer's end is also the beginning of a new cycle for bees. It is then that they prepare for the winter ahead, and their preparations, along with the help a beekeeper can give them, determine how good the next season will be.

In any part of the country where there are flower-killing frosts, bees need to store up honey for the cold months during which they will no longer be able to forage for fresh nectar. Here in the Ozarks, where winters are severe but interspersed with mild days when the bees can be active, they need about seventy-five or eighty pounds of stored honey to see them through.

In some places, bees make their winter stores principally from late-blooming goldenrod, but the Ozark bees generally scorn these flowers, preferring the *Aster ericoides*— the snow aster or Michaelmas daisy, a plant that grows widely throughout the United States. Asters, in general, are much beloved by bees; I have seen them working as happily on asters in New Hampshire or Michigan as at home in Missouri. Snow asters are tough but dainty-appearing plants with small flowers—white rays around golden centers. The foliage is delicate and feathery, reminiscent of heather, which is the meaning of its Latin species name, *erica.* Snow asters grow wherever they can gain a roothold, filling abandoned fields and edging back roads with their delicate white blossoms. They do not mind drought or light frosts, and continue to bloom bravely from August until the arrival of the first killing frost. Snow asters are so common that they are seldom noticed except by bees and beekeepers, to whom they are among the most cheerful of flowers.

I can tell when the bees have started working asters, because the nectar they gather from them is rank, and I can smell the hives a long way off. The odor so struck me the first autumn I was keeping bees that I thought perhaps there was American Foulbrood, a deadly bacterial disease, in my hives. I had never yet been around a hive with this disease, but I had read that it could be detected by its unpleasant odor. Now that I know better, the smell of aster honey does not seem bad; it is a strong, fine scent, a sign that the bees will winter well.

I like to go out and check all the hives once before winter, and do so when it is still warm enough to open them for inspection if I need to. First, though, I suit up in bee coveralls. These are made of loose-cut white cotton, with zippers in all the right places to keep bees out. They are extra long in the legs so that they can be tucked into a pair of high-top work boots, and have a zipper around the shoulders that mates with the zipper on the bottom of the bee veil, which, in turn, is fitted to the crown of a lightweight helmet with an elastic band. I wear bee coveralls whenever I work my bees, and they are a good investment for any beginning beekeeper. Those new to bees are usually nervous about being stung, and the best way to avoid being stung is to relax and move easily and confidently among them. There is nothing that gives a person more confidence in the presence of bees than to be zipped snugly inside a bee suit.

I take along a few extra beehive parts to replace broken ones I noticed when I last visited the hives, as well as some two-foot lengths of board to put under the hives if I need to replace rotting ones. Dampness harms bees, and a few boards placed there allows air to circulate and keeps them dry. I also put in the back of my pickup a tall metal five-gallon can with the top removed to hold the bee smoker I use to quiet the bees—and the tools I'll need to open the hives if I have to. I throw in a feed sack stuffed with baling twine to use as smoker fuel, some matches, my bee recordkeeping book, a pencil, my bee veil and helmet, long leather bee gauntlets, my lunch and a thermos of ice water. I am ready to go.

The first group of beeyards I am going to visit this autumn is thirty miles to the south, near one of the prettiest towns I know—

though "town" is perhaps too grand a word to describe the grocery store, gas station and cluster of houses, each with a neatly kept vegetable garden, flowers and big trees. The place is squeezed in between two ranches, thousands of acres in extent.

The ranch to the east is so large that the two beeyards I have on it are five miles apart. It is a well-managed farm. Cattle and timber thrive there and the whole setup looks as though it were an illustration from an agricultural-school textbook. Yet for a variety of reasons the bees I have on this place are not particularly productive. I am selling off beehives as I gradually whittle down my operation to the one hundred hives I want to run, and these hives will be the ones I sell next.

Directly to the west, on the other ranch, not even five miles away, is one of my most productive beeyards; I have thirteen hives there this year, and it is where I am going first. I turn right past the grocery store, drive down a gravel road through the ranch and then stop to open a gate to the lane leading to my hives. I drive through, stop the pickup and close the gate. The first rule of country living is to leave gates the way one finds them: open when they are open, closed when they are closed.

The pasture has cows in it, and so I have my beeyard fenced against them. It is a minimal fence: wooden posts that I drove with some difficulty into the rocky ground, strung around with three strands of barbed wire. It takes less to keep cows out than in, and this fence is enough to protect the beehives from the cows, who like to rub against them. In the summer, when the bees are active, they will sting the cows and drive them off, but during the winter the bees become sluggish from cold and cannot defend themselves; the cows can knock over the beehives, or at the very least push off one of the covers. When that happens, the bees inside will chill and die.

The thirteen colonies of bees in this yard, like all my others, live inside two stacked hive bodies. People new to beekeeping tell me that one of the most confusing aspects of the craft is the vocabulary beekeepers use. "Hive bodies" are sometimes called "full-depth supers," which is even more puzzling. A hive body is the basic unit of the beehive, standard at $9\frac{9}{16}$ inches by $19\frac{13}{16}$ inches long by $16\frac{1}{4}$ inches wide. It is made of clear pine, cut with dovetailed corners to make it sturdy. Indentations are cut out of each side to serve as handgrips. I always drill a hole the size of a quarter in the front of each hive body for ventilation.

These hive bodies are often called "ten-frame hive bodies," because it is customary to start new units with wooden frames that hold ten thin foundation sheets of pure beeswax, each one imprinted in a beeswax-processing factory with the matrix of the honeycomb cell that the bees would make if left on their own. On this foundation of wax, dimpled with the hexagonal honeycomb-cell pattern, the bees will build up deeper cells by adding to their foundation more wax secreted from their own bodies. Although the hexagonal cells approximate a circle, they fill the space without leaving gaps, maximizing the inside area relative to the supporting walls. This allows the bees the greatest possible volume for storing honey with the minimum of building.

These delicate wax foundation sheets are held in place by ten wooden frames suspended from the inside rabbeted front and back edges of the hive body, and may be removed easily if necessary. Before Lorenzo Langstroth, a nineteenth-century East Coast beekeeper, invented movable frames, beehives came in various forms and the bees were permitted to build permanent combs inside them. Unfortunately this meant that any work done by a beekeeper necessitated the destruction of the combs and the cruel, wholesale killing of the bees.

The impetus for Langstroth's invention of the movable frame was his discovery of what has come to be called the "bee space." Langstroth observed that bees always leave a gap of between a quarter to three-eighths of an inch between the combs they build on their own. This allows them to work on the combs and move about

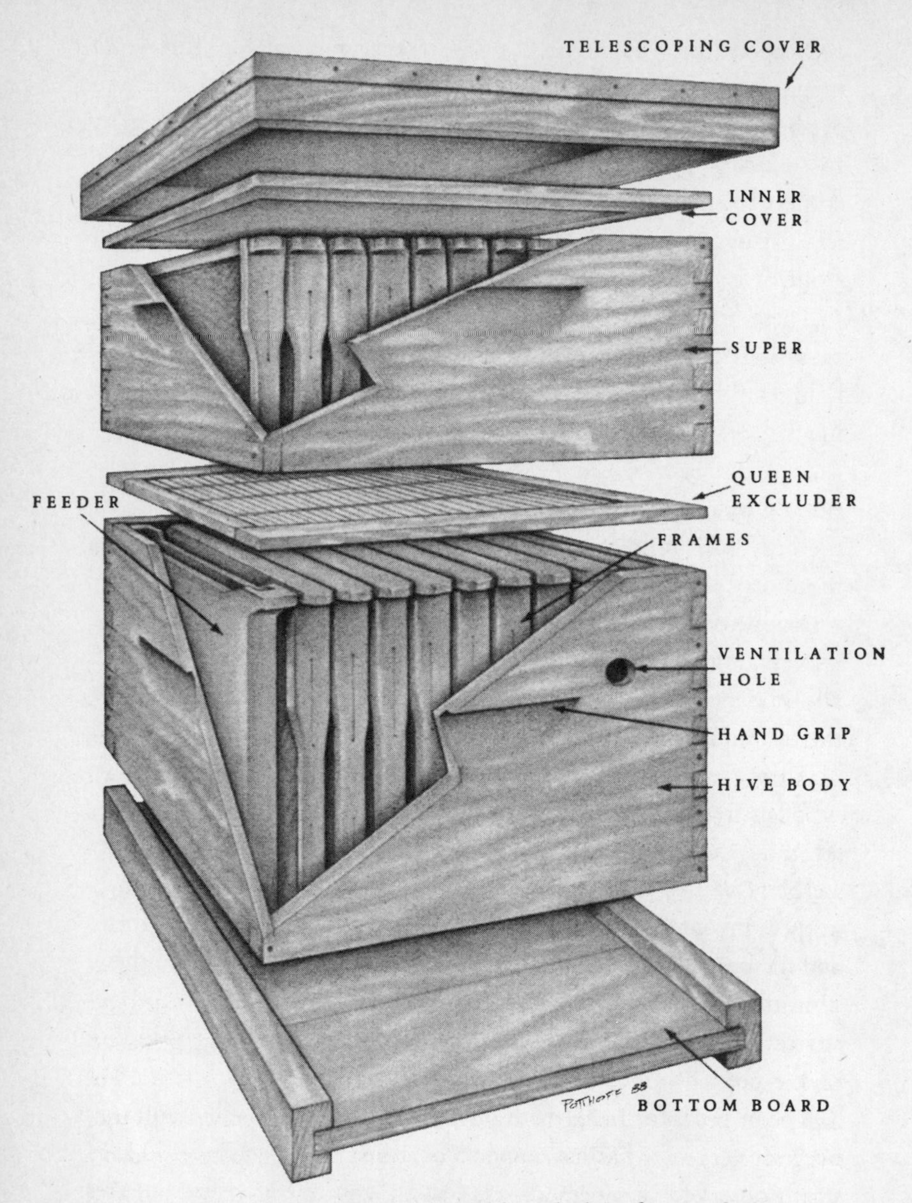

Beehive showing bottom board, hive bodies, frames, feeder, inner cover and telescoping cover

Frame held in frame grip

freely. They fill up spaces that are closer or wider than the "bee spaces" with interconnecting bridge comb. As long as the spaces between the frames of the hive are kept at approximately half an inch—a distance beekeepers quickly learn to judge by eye—the bees do not connect them with comb and the frames can be removed easily by the beekeeper with that handiest of beekeeping tools, a metal grasping device called a "frame grip."

Lorenzo Langstroth interests me. He was not only a careful observer of bees, a man clever enough to invent the modern beehive, but also the author of a gracefully written and instructive book on beekeeping, *The Hive and the Honeybee,* first published in 1853. He imported and developed the strain of bees most beekeepers still use, the Italian race of *Apis mellifera,* the sweet bee. That he could do all this in one lifetime and yet be, by his own admission, mentally unbalanced for one half of it, has always struck me as an extraordinary and admirable example of human strength.

Langstroth, who was born on Christmas Day in 1810, was brought up in a conventional family and went to Yale, where he cooperated (participated is too strong a word) with what may have been the first college food riot, the 1828 Bread and Butter Rebellion. ("The bread was not always sweet nor the butter fresh," he wrote in his *Reminiscences.*) He was later racked with guilt over his own minimal role in the ruckus: he had promised his mother to be good. He went on to study for the ministry, but upon attempting to preach his first sermon he was seized with hysterical voicelessness and was unable to deliver it. This was the start of what he called his "head troubles." He turned to beekeeping as a health-giving outdoor occupation, but when he was suffering from one of his spells of melancholia, he could not even bear to sit within sight of his hives, nor even look at the letter "B." He braved his way through his depressions by working chess problems until his mind cleared.

Langstroth was a prolific writer and journal keeper. His journals, which could be the key to an understanding of the man, may be seen at Cornell University, but present a formidable problem to the reader. Langstroth's handwriting was so bad that during his lifetime his wife, from whom he lived apart, was the only person able to read it, and it was she who put his manuscripts into readable form. In addition to their illegibility, the journals are full of private codes, mirror writing and obscene symbols. Putting them into readable English has intrigued a number of people, including me. When I proposed the project to a publisher who specialized in beekeeping and related subjects, the editor stipulated, "There are some areas in the Langstroth journals, such as the ones dealing with his periodic bouts of depression, that we would just as soon see eliminated from any publication." But it was this area that fascinated me: how a man so divided against himself could nevertheless contribute more useful knowledge and craft to the world in what was functionally only half a life than the rest of us, who are presumably in mental good health, do in the whole of our more ordinary lives. I dropped the proposal, and the Langstroth journals remain a scholarly challenge to someone with the patience, interest—and independent backing—to unlock them.

Once the bees have worked on the sheets of beeswax foundation in the ten frames and built full, fat combs in which they will raise young bees or store nectar and pollen, it is usual to remove one of the frames in order that the remaining nine can be handled by the beekeeper more easily while still making sure that the equally spaced distance between them does not violate the bees' fussy and meticulous assessment of the bee space. It is for this reason that the standard ten-frame hive body, when it is in working order, usually contains only nine frames, which makes newcomers to the craft doubt a beekeeper's ability to count.

I remove yet another frame from the bottom hive body, and put in its place on the extreme left a plastic feeder trough corresponding exactly to the dimensions of a drawn frame of comb but with an opening at the top. This is so that if I must feed the bees, I have a convenient place in which to pour the sugar syrup.

The two stacked hive bodies, one directly on top of the other, both rest on a wooden bottom board cut to the width of the hive body but with a few extra inches in front. Bottom boards are edged with a ridge around three sides that elevates the hive bodies, providing a space under the frames for the bees to walk about. Its open edge, where the projecting inches of bottom board face the front, make an entrance with a small porch; the bees use this as an alighting platform when they fly into the hive. I prefer to use cypress lumber for these bottom boards, as it is rot resistant. Although I always put hives up on boards to keep them off the damp ground, bottom boards made of pine tend to rot away even when they are treated with creosote.

This unit—bottom board and two stacked hive bodies—is topped with a flat wooden inner cover cut to the dimensions of the top of the hive body. It has a center hole to allow moisture to escape from the beehive beneath, and is overtopped with what beekeepers call a "telescoping cover." Almost all beekeepers, except for migratory ones, use this type of cover on their hives. Telescoping covers are cut slightly larger than the width and length of the hive body, and are rimmed with an overhanging edge that keeps them in place in windy weather. A fitted sheet of galvanized metal protects them from rain and snow and makes them last longer. Beekeeping supply companies ship all hive and frame parts knocked down; they must be assembled and nailed together, so a beekeeper learns to be something of a carpenter, too.

The first thing to do in a beeyard after zipping oneself into a bee suit is to take the bee smoker from its metal carrying can and light it. A bee smoker is a cylindrical firebox, compact enough to hold easily in one hand, with a small bellows attached to its edge and a hinged, cone-shaped top with a hole in it. Fuel is put in through the top opening, and the air is forced across it after it is lighted and smoldering by squeezing the bellows. This directs the smoke out of the hole in the top in puffs that will quiet the bees. The smoke makes them act as though their hives were on fire: they forget about tending their home and prepare to abandon it by filling themselves with honey; when their bodies are filled, it is difficult for the bees to bend themselves into a stinging posture.

Keeping a fire burning in a bee smoker might seem tricky to a novice beekeeper, but it really is not hard. I scrape out whatever charred fuel remains from the last time I used the smoker, then wad up a half sheet of newspaper, stick it inside and light it. When the paper is burning smartly, I put in a few lengths of loosely rolled baling twine and force air through the fire with the bellows, adding more twine before closing the top of the smoker and stuffing a wad of it in my pocket to add later. Once the fuel is burning thoroughly, the smoker may be filled all the way to the top and will smolder for hours without needing to be refueled; it will produce the thick, cool smoke that is best and gentlest for the bees. Hot smoke with sparks flying from it will incinerate them.

I use baling twine for fuel because a cattle farmer from whom I rent a beeyard takes it from the hay bales he feeds his animals in winter, stuffs it into feed sacks and loads my pickup full of it every spring. It is handy to carry around and burns well. A variety of materials may be used to fuel a smoker, though. Strips of denim or other heavyweight cotton are good, but they must be one hundred percent cotton: artificial fibers produce a foul stench and their smoke makes the bees cross. Dried grass, autumn leaves, pine needles and dried clusters of sumac fruits also make splendid smoker fuel.

When my smoker is burning well, I pull on my leather bee

Smoker

gauntlets and pick up the two tools I may need—the frame grips and my hive tool. "Hive tool" is the beekeeper's name for a small prybar, a jimmy. If there is anything bees hate it is a draft, so they weld shut the cracks between the hive parts with a gummy substance called propolis or bee glue, which they manufacture from plant resins. When a beekeeper opens a hive, he must use a hive tool to break this carefully built glue seal. One of the reasons I like to make this inspection trip to my beeyards early in the autumn is that I want to give the bees plenty of time to seal themselves back up before winter's cold begins. When I open a hive, all their meticulously crafted seals are broken, and they will have to make more propolis and chink up their hives again to preserve the warmth they generate in cold weather.

Hive tool

I don't expect to have to open many hives in this particular yard. Year after year, the bees have wintered well here because they have gathered so much aster honey. I smelled it as soon as I stepped out of the pickup, and when I walk toward the hives, I can see streams of bees flying purposefully in and out of their entrances, a sign that they are hard at work on a nectar flow. When only a few flowers are blooming, the bees fly around in a desultory way, and often, if the weather is warm, they hang aimlessly on the front of the hive or stand in bunches on the alighting board. Today, the bees are active; the ones returning to their hives fly low, heavy with nectar. A quick check of the front of the row of hives shows only one that may be in trouble. A heap of fine, powdery beeswax obstructs part of the entrance. I make a mental note to open that hive and check it. Beehives should always be opened and worked from their rear in order not to intrude on the bees' flight path. Bees are singleminded and fly directly to their destination. Today they are flying straight to the field of snow asters I can see across the river branch. If I were to stand in their way, it would divert and annoy them; they might sting.

I move behind the row of hives and put my hand in the handgrip of the bottom hive body on the first hive in the row, hefting it to estimate its weight. It is heavy with honey; its bees should winter well. I have made a note to replace a broken inner cover on the second hive, so I pick up the smoker and puff a stream of smoke into the hive entrance to subdue the guard bees posted there and prevent them from spreading an alarm. With my hive tool I loosen the telescoping cover and take it off, prying up the broken inner cover and puffing more smoke into the top of the hive, which is filled with bees in bustling numbers. They have felt the jarring disturbance, and their abdomens are raised defensively, ready to sting if necessary. The smoke, however, creates a more immediate concern, and they scurry down between the frames to avoid it. I toss the broken inner cover in the direction of the pickup, replace it with a new one and close the hive.

I work my way down the row of hives, hefting them and assuring myself that they are all in good shape. One needs a few boards under it to maintain an air space from the ground. Carefully, without bothering to smoke the bees, I lift up the front and back edges of the hive and push the boards under it. When I come to the hive I thought might be in trouble, I find that it is much lighter than the others and its honey stores are low. There are not many bees inside, and the ones that are there are clustered over only a few frames in the upper hive body. At this time of year, a normal bee colony is still close to its summertime strength of about 60,000 bees. There can't be even a quarter of this number here. I loosen some of the frames with my hive tool and use my frame grips to pull it out. The frame has recently had honey but now it contains no honey at all. The ragged edges of the wax sealing the honeycomb show that the cells have been ripped open hurriedly and carelessly to get at the honey stored inside. This is a colony of bees too weak to defend itself from other bees, and it has been robbed of its honey by more spirited neighbors. The powdery wax I had spotted at the hive entrance earlier meant that the honeycomb had been opened by intruders, for in a strong, well-regulated colony the bees never leave any debris in their own hive.

A check of the rest of the frames shows the bees to be free of disease, although there are signs that wax moth larvae and cockroaches, opportunists that they are, have begun to take advantage of a hive of bees too few in number and too poor in morale to defend itself from enemies. The nondescript, medium-sized gray wax moths lay their eggs in beehives whenever they can, and after hatching, their larvae—wriggling dirty-white caterpillars—feed on honeycombs, destroying them in the process if the bees are unable to kill them first.

"I had some bees once," I hear from now-and-again beekeepers, "but the moths got 'em." This is confusing the sign with its cause. Wax moths are everywhere. A strong bee colony can defend itself from their damage, but once there is a drop in strength and morale the balance will tip in favor of the wax moths.

As I check the frames of this hive, I find there are a few bee larvae being raised by worker bees. This indicates that they still have a queen, the single fertile female bee each colony requires. I check my record book and find this was one of my most productive hives the previous year, and that early in August I had taken six honey supers from it. "Honey supers" are about half the height of a regular hive body, and stack on top of the hives. In them the bees store the extra honey that a beekeeper may safely remove—this is his honey crop. Because honey supers are smaller than hive bodies, they are lighter and easier to handle, easier to carry empty to the hives in springtime and easier to remove filled at the end of summer.

The six supers I took from this colony are, on the average, twice the number an ordinary colony will fill with honey. So a few months ago this colony contained a greater number of bees than an average colony does. There may have been 100,000 or more only recently. What happened to them?

Over the years, I have discovered that certain bee colonies—particularly prosperous hardworking ones like this one had been—sometimes feel impossibly crowded when their supers are taken away and they are reduced to their basic two hive bodies. This is one of the reasons why they swarm—that is, split a parent colony into two and fly off with the old queen to take up new quarters in a hollow tree or under the siding of a building. Bees usually swarm in the spring, but occasionally I see swarms in the autumn, after the supers have been removed. This is not good survival strategy for the bees. The autumn swarm that buds off from the original colony will not have enough time to set up a new, well-stocked household before the frosts come and kill the flowers. And the parent colony, even though it has raised a new queen, is usually too weak to defend itself against robber bees and contains too few bees to winter well. This is probably what has happened with this colony.

If I had just a few hives rather than three hundred, this late swarming would be easy to prevent. If I put an empty super with ten frames of foundation on a hive when I remove the filled honey supers, the bees would have enough space and enough to do to prevent the swarming impulse from developing. But the additional expense of keeping three hundred foundation-filled supers on hand, and the labor of putting them on and taking them off just to prevent a few swarms doesn't make sense from a commercial beekeeping standpoint. However, if I leave these bees as they are, they will surely die over the winter. Their unprotected combs will be destroyed by wax moth larvae. The beekeepers' maxim—take your winter losses in the fall—is a sound one. I shall add these bees to another hive, and use the old beehive parts to make a new replacement hive in the spring.

"Adding the bees to another hive" sounds simple. It isn't. If I just dumped these bees in with others, they would all be killed as intruders. Every hive is made up of three castes: the drones, the workers and the queen. The drones are the male bees, present only in the spring and of no concern in the fall, when all but one of the bees in each of the colonies are members of the worker-bee caste. Workers are females with atrophied sexual characteristics, and they do nearly all the work in the hive. They gather nectar and pollen, make propolis, raise young bees, build comb, make honey, defend the hive and take care of the queen. The queen is the single representative in each hive of the third caste of bees. She is a fertile female, mother to all the bees in the colony. By being constantly touched and fed by her daughters she spreads her own special chemical marker—a pheromone—throughout the colony. It is this that makes each colony unique and distinct from any other. Queen bees are jealous and cannot bear the presence of another queen. When two queens meet, they fight to the death. And if the offspring of two different queens are abruptly forced together, they will do the same.

Before uniting this weaker colony with a stronger one, I must find the queen of the poorer colony and kill her. I do not want to give her the chance to kill the queen in the prospering hive where I shall house her bees. The queen, a long, elegant bee, is easy to find among the shorter, stubbier workers, especially in such a small colony. I find her quickly, and—regretfully, because I do not like to harm bees—I kill her with the sharp end of my hive tool.

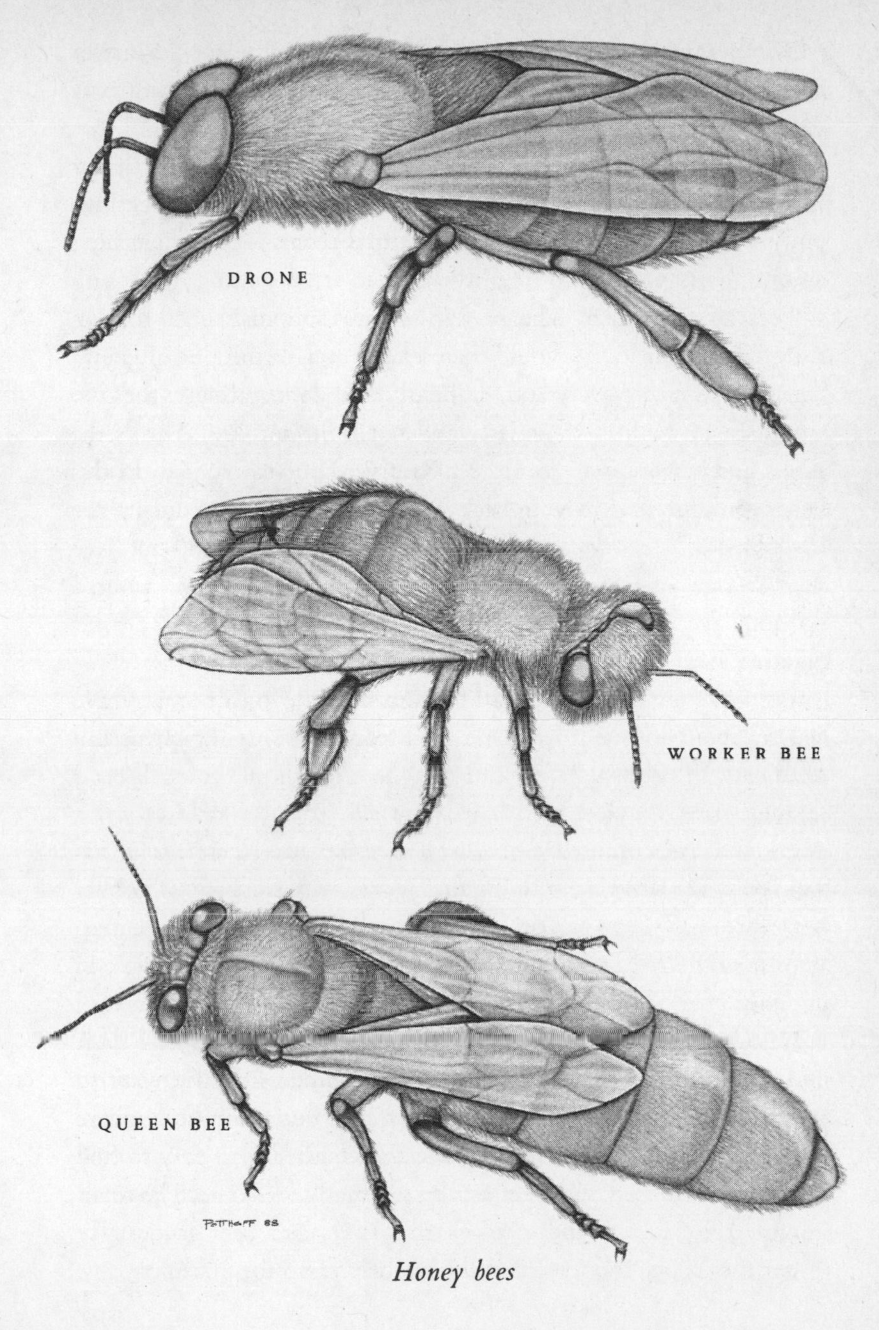

Honey bees

The bees from the weaker colony, however, still carry the pheromone from their late queen. Until it fades, they will be instantly recognizable as foreigners by the bees with whom I want to join them. If I were at home and could return easily to these particular hives, I should unite the two colonies by putting the upper hive body—the only part of the hive that contains bees—of the queenless colony directly on top of the hive next to it, separating them with a sheet of newspaper. The bees, fussy in their dislike of foreign material in their hives, would then set themselves the task of chewing the newspaper away and, gradually and cooperatively, the two sets of bees would meet quietly and not fight and kill. This is the surest and kindest way to unite two hives. But to try and do this here would involve driving back in a week's time and reducing the combined three-story colony to two so that the bees would not have such a large interior space to keep warm throughout the winter. I do not have the time, and I take a shortcut, one of many I must practice as a commercial beekeeper.

I remove the hive bodies and bottom board of the queenless hive and set it off to one side. Some bees from it fly up, agitated, and return to the original hive location, where they join the small band of their sister foragers who have returned from the field of snow asters and are confused and disoriented because their home has vanished. Tentatively, without the aggressive assurance of robber bees, they drift a few at a time into the two hives that had once been their neighbors. I stand and watch. Guard bees are on the alert in the entrances of both those hives, aware that something untoward is taking place. They examine the incoming bees, letting pass their own foragers but checking the drifters suspiciously. They fight a few to the floor of the hive and kill them, but most are allowed through unharmed. Returning foragers, no matter what colony they are from, are loaded with nectar and are usually welcome. The other stragglers are so few and so meek that the guard bees grudgingly accept them. The frames that still have bees from the queenless hive clinging to them I rap sharply against a tree. The dislodged bees fly up in distress, circle and try to get their bearings. They, too, eventually return to the original hive location and drift into the neighboring hives; they are so demoralized and unaggressive that the guard bees accept them, too.

I heft the remaining hives and find them all in good shape. I put heavy rocks on top of each one to keep the telescoping covers secure through autumn and winter storms, then fill out my record book, noting the date, what I have done for each hive and my assessment of their honey stores. I put the smoker back in its carrying can, where it will smolder safely until I get to the next beeyard; the hive tool and frame grips go on top of it, and I'm ready to leave. I probably won't visit this beeyard again until next March.

Many beekeepers recommend reducing the entrance to the hives in the fall, and years ago I used to do this. "Entrance reducers" are thin blocks of wood cut to the length and height of the hive entrance, notched with a hole small enough to keep out mice but large enough for a bee to pass through. Mice are destructive to beehives during winter. The inside of a hive is a warm, snug place; during cold weather, when the bees are sluggish, the mice like to go in and chew up combs, make their nests and raise babies. The entrance reducers prevent them from getting in. But when I did reduce my beehive entrances in the winter, I often found that by spring the tiny entrance notch had become blocked with dead bees. Bees in every colony die from old age during the winter, but with the entrance reducers in place, the remaining bees—which, of course, could fly through the ventilation holes in the hive—had not been able to carry away their dead sisters as they would have done had the entrance been unblocked. Under normal circumstances, bees are very mindful of hive sanitation but these piles of dead bees were damp, moldy, sour and unwholesome.

One autumn, after having harvested a record amount of honey, I was too busy trucking around the country selling it to visit the hives and put the entrance reducers in place. Next spring, although there was some mouse damage to the combs, all the hives were dry, sweet and clean, and the bees were in the best health I'd ever seen them. After that, I stopped reducing the entrances, and now I accept winter mouse damage as a tradeoff for better ventilation and healthier hives.

The store in the little town between the two ranches carries more than groceries. In back of the shelves of peanut butter and canned corn are work clothes, axe handles, coils of rope and other country necessities. In the middle of it all is a Formica-topped table, where ranch hands are usually gathered drinking coffee from the pot that sits on a hot plate at the end of the meat counter.

"Howdy, Bee Lady," one of the hands calls to me, and another shoves out a free chair for me to sit down.

I pour a cup of coffee and sit with them. I don't know their names, but they know mine: Bee Lady. A middle-aged woman in baggy white coveralls who smells of burnt baling twine is a standout in any crowd.

I can't think why it is so few women keep bees. Among commercial beekeepers—those with three hundred hives or more who make their living from them—there are few to be found anywhere. I know I am the only one in my area, and I may be the only one in Missouri. Even among those who keep just a small number of hives, few are women although there are a number of women entomologists who specialize in bees and who work for the bee research laboratories.

I should like to think we have changed since the days of Cyula Linswik, a woman beekeeper who could, in 1875, advise women beekeepers against making their own frames. She wrote:

> Let her spare her gentle fingers for finer uses—as sewing on of buttons—and buy the frames. . . . One of the thorns

> in the path of the woman who undertakes to master the theory and practice of bee-keeping is her lack of natural or acquired ability to drive a nail straight, to use a saw with safety to the implement, or a sharp knife with safety to herself. The gifted few of whom this may not be true constitute so small a fractional part of womankind that they may be regarded, properly, as exceptions proving the rule. And the woman who begins to keep bees without having her attention directed to this matter is in danger of suffering from vexation of spirit and wounded fingers many times during the course of her novitiate.

Two years later the same lady observed:

> Apiculture, like most out-door avocations, is almost monopolized by the stronger sex. In the days of our grandmothers this was a natural and necessary consequence of man's fitness and woman's want of fitness for the work. Picture a woman's helplessness in view of a swarm safely clustered in the top of a tall tree! Imagine her lighting the brimstone and piteously dooming to death her faithful little laborers—if you can. Need we wonder then that ere the introduction of movable frames women did not aspire to be bee-keepers? But that so few women are interested in apiculture today is less easily explained. . . . Does apiculture offer any special inducement to women? May it not be that the work, no longer impossible, is still for them undesirable?

I don't know the answer to Linswik's questions, but some things have not changed very much. One of today's respected beekeeping encyclopedias, *The ABC and XYZ of Bee Culture,* has an entry under "Beekeeping for Women" which is as insulting and patronizing in tone, intent and phrasing as Linswik's writing. There is, of course, no entry under "Beekeeping for Men," nor should there be. The keeping of bees has nothing to do with sex, but I am reminded of Samuel Johnson's observation that the matter of a woman's ability to preach is similar to that of a dog's ability to walk on his hind legs: the wonder is not that he does it well, but that he does it at all.

The ranch hands and I started talking weather because it is of real concern to all of us. It determines their hay crop and my honey crop. In 1980, a year of searing drought, I was able to harvest only six thousand pounds of honey from my three hundred hives and lost bees to starvation in midsummer. In another year, when the rains came at the right time, those same three hundred hives gave me thirty-three thousand pounds. So we talk weather. We talk hay. We talk bees. We talk farm prices and shake our heads sadly.

Fifteen years ago, I came to this part of the country with nothing to recommend myself. I had been a university librarian on the East Coast. Local people are slow to accept newcomers; they have seen many of them come and go, and I am sure they thought I was one of those bookish types with a head stuffed full of theories about how I could live in the country better than people born to it. But I am not good at theory, and I certainly wasn't going to spin any about cow or pig farming. Instead, the word went around I was trying to make a living at bee farming. No one had ever done that here. Bee farming, it turned out, was different from cow or pig farming, but not all that different. I work hard and sweat a lot. So do they. None of us is making much money, but we stay with farming because we enjoy it and like to mess around with animals.

I cannot sit too long talking, because the empty hive body I took from the beeyard is in my pickup, which is parked out in front. Soon, bees will be attracted to it, and I don't want to take the chance of anyone being stung. I get up to leave.

"Ought to be getting back to work, too," the foreman of the ranch where I have just been remarks. "But I figure if I pour myself another cup of coffee the mood will pass."

I pay for my coffee and go. Work at the next yard is routine, and I finish up quickly there. This yard is on the second ranch, but the bees are close enough to town to benefit from the flowers and gardens people have there and the clover growing in their lawns. The hives are already heavy with honey for the winter.

At the third yard, the last one in this group and the second one on this ranch, I am going to prepare the hives for moving. The bees here are the least productive of any I have. The colonies build up quickly in the springtime, better than in any other yard, making it seem as though they are going to be record-breaking hives. In the woods nearby, there are many wild fruit trees and serviceberry, which provide them with an early source of pollen and nectar. But they never live up to their promise, and their summertime production is disappointing. The ranch is well-kept from a cattle-and-timber standpoint, but the bees would be better off if it were not. The fence rows are kept clean of weeds, many of which would be good sources of nectar for bees. There are no overgrown, unused pastures filled with blackberries, straggling sumac, wild mint, sweet clover and other wildflowers which the bees would know how to use. Instead, the fields are lush with sensible fescue, a grass that stays green nearly the year around and provides pasture for cattle. But a field full of fescue is no better than a desert for bees. Acres of alfalfa have been planted on this ranch for hay, and alfalfa blossoms produce a nectar that makes superior honey, but the ranch hands cut the hay on such an efficient schedule that it seldom reaches the blooming stage, so the bees have no benefit from it.

Today I want to check the bees to make sure they are healthy and strong and prepare them for moving. They are all in good shape, I find. I replace one telescoping cover that is beginning to dry-rot, and set about readying these hives for moving.

The yard can sustain only eight hives, and it does not make economic sense to keep a small beeyard so far from home. Not long ago, a man telephoned and asked if I would sell a few hives. I gave him a good price for the eight in this yard, and tomorrow I am going to meet him here at the end of the day, after the bees have returned from their final foraging flights, to help load the hives on his pickup.

Parts of the beehives are sealed together with propolis, to be sure, but if the hives are loaded into a pickup and driven over rough roads they will jiggle apart and the bees will escape. The bees will be angry over the disturbance to their hives and will sting whoever is nearby, which would add to the difficulty of moving them. But, beyond that, those who escape will be lost, and, separated from the intensely social community that their colony is, they will die.

Bees have a keen and precise sense of place. When they fly out of their hives, they commit to their memories an exact picture of all the significant landmarks near it. It is such a careful picture that if their hive is moved even ten feet away, their home is as good as lost to the returning foragers. And because their map of a foraging area—perhaps five square miles—is so accurate, a beekeeper who wants to relocate a given beeyard in the same general area must first move the bees at least ten or fifteen miles away and leave them for a week or so until they forget the map of the original location by learning a new one. After that, he can move them back to a spot he prefers near the old location. But if he were to move them directly there, they would stubbornly fly back to their old home place.

To keep the bees from escaping during tomorrow's move, I am going to fasten together all the hive parts and block all the holes. I have with me a box of hive staples—copperplated flat wire staples, two inches long with ¾-inch ends. I drive in a pair of them, attaching the hive body to the bottom board on one side, and repeat the process on the other. Another pair on each side holds the two hive bodies together. I drive each staple at an angle to its partner, so that the hive bodies will not shift.

I am using a hammer, and the pounding disturbs the bees, even though I have smoked them before I started the job. They come out to investigate the fuss, and I give them an extra puff of smoke to quiet them. That does not prevent one from finding a place on the

Hive staple and moving screen: hive is shown ready to move, with staples and screen in place

back of my neck where the veil is lying directly on my skin. She stings me soundly through the veil to express her dislike of what I am doing. I have been stung enough so I have no reaction to bee venom. There will be no redness or swelling on the back of my neck. Tomorrow, the spot will not itch. But I can feel the sharp prick of her stinger as it goes in. It is no worse than being snagged by a blackberry thorn. I am seldom stung, not only because I am used to the bees and relaxed as I work with them, but because I don't often do things that make them cross, such as pounding on their hives with a hammer. I long ago gave up a number of beekeeping practices conceived with the notion of making bees do certain things that seemed good from a human standpoint but which usually involved radically disrupting the hive. Instead, I watch the bees more, try to understand what they are doing and then see if I can work in a way that will be in keeping with their biology and behavior. I try to create conditions that will make them happy, and then leave them alone as much as possible. Fewer disruptions allow them to produce more honey—my per-hive yields are greater now than in my first years of beekeeping—and they don't often have an occasion to object to my presence, so they don't often sting me.

Like many beekeepers, I have discovered a dose of bee venom from a sting alleviates the symptoms of arthritis, and I am stung so infrequently that when the joints in my hands begin to ache I have to go capture a bee and force her to sting the place where the hurt is. The pain from the arthritis is gone by the next morning. This is anecdotal evidence, and as such has no value for science, but it will not stop me from taking my therapeutic stings and hoping I can keep my hands limber. By my age my mother and grandmother each had hands so crippled with arthritis that they were hampered in their daily activities.

Those new to beekeeping usually get stung a lot. I was too, when I first began working with bees. Bees' vision is such that they can easily see objects that move quickly and jerkily, and often a person unused to bees is tense and nervous when he is around them. His jumpy motions will attract their attention. Bees are myopic. Large stationary objects near them appear fuzzy and indistinct, but they are very sensitive to broken patterns, the flickering of light and sudden movement. They are quick to see flowers swaying in the wind and enemies trying to get into their hives. Bees, like other animals, are quieter when those around them are relaxed and move slowly.

Some people are seriously allergic to insect stings; they should not go near bees. They break out in welts on the soles of their feet and the palms of their hands, experience shortness of breath and accelerated heartbeat. After repeated stings, they may go into anaphylactic shock and die. Such people are very few in number, and some doctors have begun treating them with whole bee venom in an attempt to help cure their allergy.

The normal reaction to a bee sting is swelling, redness and itching. It is easy enough to desensitize anyone who reacts that way.

I usually hire a strong young man to help me with the honey harvest each year and before I let him out in the beeyards I make sure he will have little or no reaction to bee venom. I start him on a desensitization program a couple of weeks before he is to begin work. On the first day I put an ice cube on his arm to numb both it and his fears. Then I place a bee on the spot, holding her by the front part of her body so she can curl over her abdomen and sting him. Angry at the restraint, she usually obliges and then pulls away, leaving her stinger behind in the young man's arm. The bee, her stinger gone, will soon die.

The stinger is a curved barb topped with a bulbous venom sac. Muscles in the sac continue to pulsate after the bee has left, and the muscular contractions force the barb on the stinger deeper into the flesh. I want my helper to have only a partial dose of venom the first time, so after a few minutes I scrape the stinger out with my fingernail. The stung place will soon become red and start to swell.

The next day it will be itchy. On that day I repeat the process, but leave the stinger in place ten minutes, so my helper can receive a full dose of venom. I continue giving him one sting a day until he no longer has a reaction to it. Then I increase the number of daily stings to two. Again, he will react until his body stabilizes at the new level of venom. I continue this way until he can tolerate ten stings a day with little reaction; after that I do not worry about taking him out to the beeyards. Everyone reacts differently, and I never can tell exactly how long the process will take, but for most the first part takes longer than the last. Usually after four, five or six stings, the young man reaches a plateau and can move on one day at a time to the required ten.

Once he discovers that getting stung is not really painful or scary, he dispenses with the ice cube and starts administering his own stings. He takes control. He has lost his fear and will relax around the bees, which will, in and of itself, make him less likely to be stung.

After I have stapled all the hives I move around to their fronts to screen the entrances. It would not be kind to close up the hive entrances and ventilation holes with something solid because the bees inside would suffocate. So I am going to put in each entrance a screen of hardware cloth, cut to the exact width of the hive entrance and crimped into a V. I poke the point of the V into the entrance—which makes the bees, already out of sorts, crosser still. I smoke them a bit more, and then, using a hand staple gun, fasten the screen to the bottom board to prevent it from jiggling out in the drive to the bees' new home.

I have one thing left to do. From the pickup I take a roll of gray duct tape, tear off small strips and use them to block one of the ventilation holes in each hive. This still leaves the bees the other ventilation hole as an entrance and exit, and air can move through the screen in their usual entrance. Tomorrow, when I meet the buyer of these hives here at dusk, all that will remain to be done is to tape the second ventilation hole in each hive and we will be ready to load them on his pickup.

Moving beehives is a two-person job. This time of year, of course, the hives are heavy with honey: their total weight may be two hundred pounds or more. But a two-story beehive is impossibly cumbersome for one person to handle alone at any time. It takes two people, one on each side, to lift up a beehive and carry it to a pickup. I'll help the new beekeeper load these hives, and when he gets home his son will help him unload them.

Years ago, when I first began keeping bees, I wanted to increase the number of hives I was running from the sixty with which I had started, and do it as fast as possible. I bought a hundred hives that were tucked away in groups of six and seven in remote beeyards along the banks of the Missouri River. A friend helped me move them, and we rented the biggest U-Haul truck we could find in order to complete the job in two trips.

We brought along my pickup, as we needed it to ferry the beehives between the outyards, often at the end of the narrowest of lanes, to the big truck we had to leave parked on wider roads. In each outyard, we would screen and staple the hives and then tape all the holes we could find. But the man who had sold me the hives had rather lost interest in them; many were in need of repair and had holes in surprising places. The rotted bottom of one fell out as we lifted it, and a horde of angry bees rushed to fasten themselves on to our bee suits, stinging us through the fabric which was damp with sweat and clung to us. We moved that particular hive to the big truck as quickly as we could, but the bees stayed cross, and as we filled the U-Haul we found that bees in many of the other hives were discovering holes and cracks and chinks that we had not. They hovered around the dark interior of the truck, and then, each time we opened the big sliding door, they would fly instinctively toward the light.

The last yard from which we were going to move bees on the

trip was down an overgrown footpath. We drove the pickup as near as we could, and then carried the beehives through brambles to it. After we had loaded the last one, I backed the pickup around and drove down the twisting road to the big truck. As we rounded the final curve, we noticed there was a strange pickup parked near the U-Haul. Two men got out of it and looked around furtively, but did not see us. They tiptoed over to the truck, their curiosity piqued by an apparently abandoned U-Haul.

They tried the sliding back door gingerly, and found it would open. They gave it a push. The loose bees inside rushed out toward the light and enveloped the two men in a furious buzzing cloud. The men were both heavy, with ample beer bellies, but they ran like jackrabbits to their pickup and drove off at top speed, careening from one side of the road to the other as they tried to brush bees from their heads. I'll wager that is the last time either of them meddled with an abandoned truck.

II

THE BEEKEEPER'S WINTER

The Beekeeper's Winter

Another way to get started with bees is to order all the needed parts of the hives in the winter, assemble them, and then, in the spring, buy what are known as "package bees" to fill the new hives.

This method has some advantages and one disadvantage. Package bees, which include one queen and several pounds of bees unrelated to her, are shipped in screenwire cages through the U.S. mails. They are an artificial assemblage and the two or three days they spend in transit additionally confuses and disorients them. As a result, they are sometimes difficult to hive and may even refuse to stay in the quarters the beekeeper has prepared for them.

But new bees purchased from a reputable bee breeder will almost certainly be free of disease, and the new comb that the bees will draw out on the sheets of wax foundation will serve for many years. In contrast, established hives must be checked carefully for disease and will usually contain old combs that will need to be culled and replaced.

Some cities have stores in which to buy the tools needed for beekeeping, and the hive parts, too, but the major suppliers for beekeepers sell by mail order. There are a number of these companies, and the prices and quality of their products vary. All of them advertise in the two grand old beekeeping publications, *The American Bee Journal* (published since 1861), Hamilton, Illinois 62341, and *Gleanings in Bee Culture* (published since 1873), P.O. Box 706, Medina, Ohio 44258. In addition to serving as a display shop in print for beeware, these magazines feature useful practical articles on beekeeping techniques. But they are published by two of the major suppliers of beekeeping products and have a distinct interest in selling tools, supplies and bees as a result. Their orientation is toward what a person can buy to become a better beekeeper. I have always held there is too much emphasis on gadgetry in beekeeping. Bees are forgiving animals, and will tolerate a good deal of rearrangement in their lives for whatever new fad sweeps the beekeeping industry—plastic foundation and frames, double queen management, tar-paper wraps for winter—but the best beekeepers I know are those who let the bees themselves, not equipment manufacturers, be their teachers.

That said, a new beekeeper still needs basic equipment, and there is no better place for him to start than by asking the major advertisers in the two magazines mentioned above for their catalogs. After these have arrived, he can compare prices and shipping rates. He will need the following tools:

1 bee smoker
1 hive tool
1 pair of frame grips
1 bee veil

Gauntlet-style bee gloves, a bee suit and a bee helmet make the job easier; they will keep him from extra stings, but they are not strictly necessary. He can substitute work gloves, an ordinary hat and a second layer of work clothes securely fastened.

For each hive he is going to start, he should order:

1 bottom board
2 hive bodies
1 inner cover
1 telescoping cover with metal over-cover
20 deep frames for hive bodies
20 sheets of wax foundation to match frames
1 feeder
Nails for fastening the knocked-down hive parts together
Tinned wire for frames

Later on he can decide whether he wants to have his bees make comb honey or whether he wants to extract the honey they will produce. This decision may be deferred for a while, because package bees started on foundation do well the first season simply to set their house to rights and produce enough honey for themselves to get through the winter; they seldom produce any extra honey for the beekeeper. But the size of the supers he will order later depends on his decision about comb or extracted honey.

In my part of the country I need, on the average, three supers for every hive; since I do not produce comb honey, I use what are called medium depth Illinois supers. Comb honey supers are shallower than these.

Buying a hive of bees is, in some ways, like buying an Irish Setter puppy: it changes one's life. But having two, or even three, hives of bees is not like having two or three Irish Setter puppies. The first hive is a Big Deal. The additional ones are not. Two or three hives are no more trouble than a single one, and in fact they make the whole task simpler.

Accidents can happen to queen bees, and in a single hive the bees

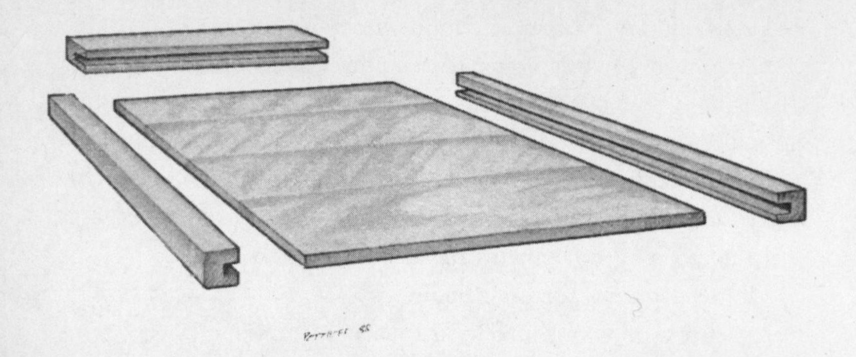

Bottom board parts

will die out if they lose their queen unless they can raise a new one quickly or have their beekeeper give them one. Ordering queens by mail may take several days, and by the time she has arrived the bees in the queenless hive are often so demoralized they will refuse to accept her. But if a "queenright" hive, one with a healthy, fully fertile laying queen in it, is sitting next door to the queenless one, the beekeeper can take a frame of eggs and young bees from it, give it to the queenless hive and, in most cases, the bees will raise a new queen for themselves. In addition, frames of honey and young bees can be interchanged if need be. This is why, when I help new beekeepers get started, I always suggest that they begin not with a single hive but with several.

On winter days, when the snow is blowing, I like to go out to my barn and build up a fire in the woodstove. After it is warm there, I spend the day putting my beekeeping equipment to rights. My introduction to carpentry was assembling beehives. It is not difficult.

The first thing to do after receiving a shipment of beehive parts is to sort out the similar pieces and see how they fit together.

Bottom boards from different manufacturers differ slightly, but most come with several grooved boards that fit into one another and that, in turn, are held in place by side rails and two strips of wood that meet the rails at the back.

The bottom boards, as well as the other hive parts, will fit tighter if they are glued in addition to being nailed, but simply because of the added time needed I do not glue mine (I should probably spend less time repairing them if I did so) but instead merely fit the bottom board parts together and nail the rails to the ends of the grooved

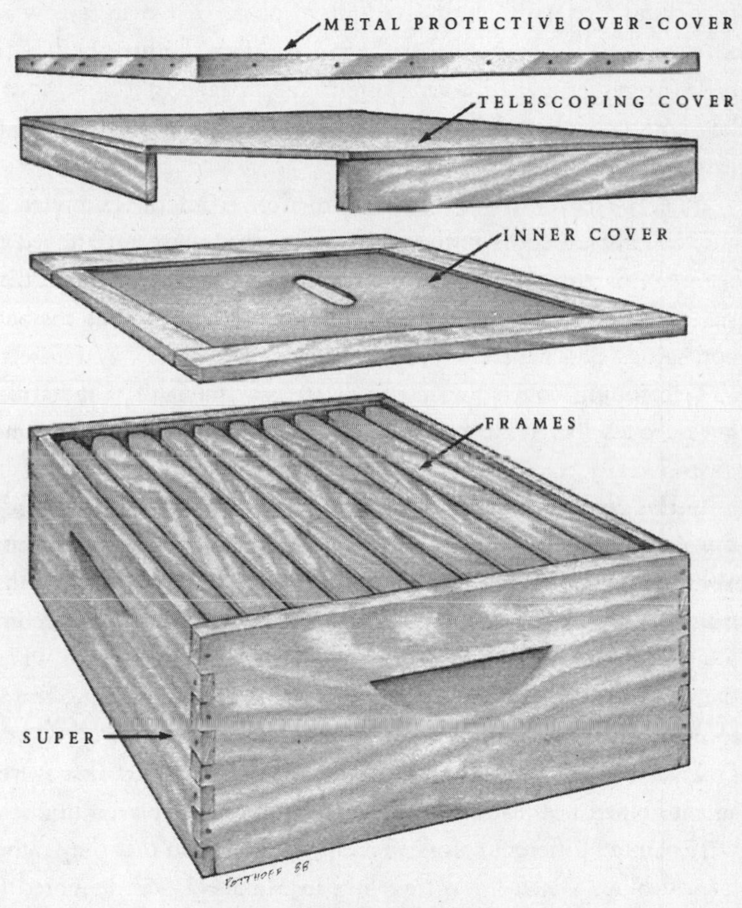

Hive body showing
inner cover parts and telescoping cover

bottom with galvanized sevenpenny nails. I treat the undersides of the bottoms and the exterior of the rails with creosote, and scatter the finished bottom boards outside to allow the chemical to weather for at least three months before using them with the bees.

It is easy to see how the hive bodies fit together with the matching dovetail ends meshing neatly. Most beekeeping supply companies also drill nail holes along the corners made by the interlocking dovetails, so it is a quick job to drive in a row of sevenpenny nails to hold them in place. After the hive bodies are nailed together, I drill a ventilation hole in each one, and, since I buy commercial grade equipment, I paint all knots with shellac to help seal them.

Inner covers also vary in construction from one supplier to another, but all come with wooden rims that must be fastened to the main part of the inner cover. Once these are on the hive, they will be pried up frequently, and often come apart unless they are both glued and nailed.

Telescoping covers also come in an assortment of manufacture but once assembled all should be additionally protected by a metal over-cover.

In this country, for reasons I have never understood, it is traditional to paint beehives an unimaginative and antiseptic white. White is reflective, and if the hives are placed in the direct sun, that may be a kindness to bees. But hives, even white ones, can become too hot in the sun and should always be set out in a place where they will receive some afternoon shade. So placed, there is no reason to paint them white if a beekeeper prefers another color. Karl von Frisch, Nobelist, zoologist and student of bee behavior, discovered that it helped bees to find their own hives if those placed in a row were painted different colors; in addition, he found that they rather fancied blue. When I was traveling in Mexico I was delighted to see beehives painted in deep vibrant colors—red, green, blue and black. Whatever the color, the exterior parts of hive bodies and the

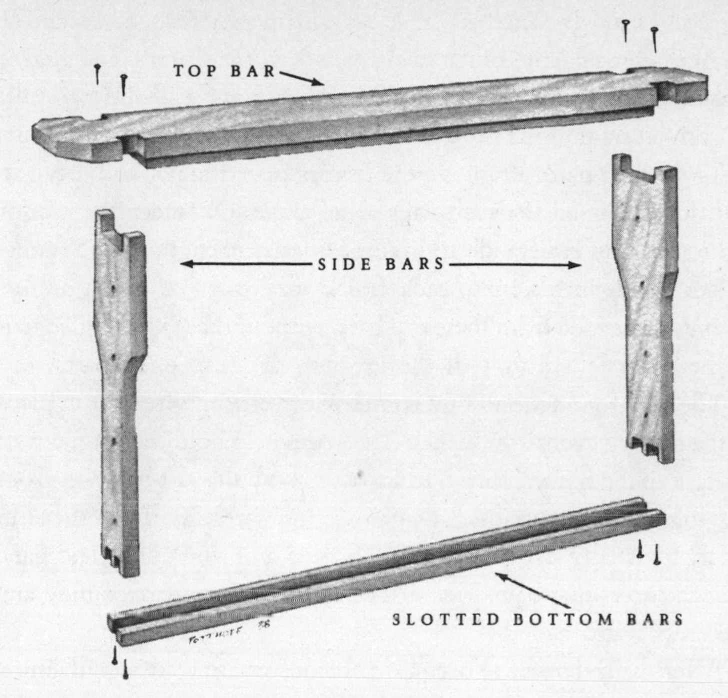

Frame parts

telescoping covers should be painted to protect them from the weather.

The frame parts, because they are so many, often look bewildering when they are first unpacked, but once they are sorted into piles it is obvious how they fit together. They include "end bars," which make the sides of the frames; "top bars," which suspend them; "wedges," which fit against the top bar to secure the wax; and "bottoms," which complete the frame. I prefer to use what are called "split-bottom" frames. The two split separate lengths of wood that make the bottom serve as a holder for the wax foundation.

I have to assemble so many frames that I have a jig, in which I can put together ten at a time. The jig holds twenty of the end bars, ten on each side, at the correct distance, while I nail the top bars in place with 1¼-inch nails. I use an upholsterer's hammer for driving these and the other small nails into the frames, and find it makes the job easier. Then I flip over the jig and nail the two split bottom pieces into the grooves cut into the bottoms of the side bars, using ⅞-inch nails. I remove the frames from the jig, and drive an additional pair of 1¼-inch nails at an angle just under the ears of the top bar on either side to give additional strength to the frame. When filled with honey, each frame may carry as much as five pounds suspended from these top bars; without that extra angled nail at the point where stress is the greatest, they can pull apart.

The wax foundation I buy comes with vertical wires put in place at the factory every 1¾ inches. The wires help hold the fragile wax straight in the frames. But in addition, to keep the soft beeswax from sagging in warm weather, I put in cross wires. To hold those in place, I partially drive two ½-inch nails into one of the end bars, near the holes that have been drilled at the factory to accommodate the cross wire.

After the frames have been assembled they are ready for the wax foundation. The wax is brittle in cold weather, and shatters easily, so I always wait for the barn to be warm before I handle it.

I rest each frame on a special frame holder as I insert the wax and wire it, but for assembling just a few a board cut to the inside dimensions of a single frame would be enough to make a level surface for working with the sheet of wax. There are small hooks, extensions of the vertical wires, on each sheet of wax, and these are placed at the top of the frame after the foundation sheet is slipped between the split bottom. The wedge, the last remaining frame part, fits snugly against the top bar and holds the wire hooks in place. Seven-eighths-inch nails driven at an angle will keep the wedge secure.

All that remains to be done is to cross-wire the wax foundation sheet to the frame. I have a commercial wiring device that makes the job easier when working with a number of frames, but for assembling just a few its purchase is unnecessary. A homemade holder for the roll of tinned wire would do just as well. A length of wire is pulled through, starting on the side where the ½-inch nails have been driven. One end of the wire is secured to the top nail, which is then driven home. The wire is threaded across the frame through the hole on the opposite end bar, back again through the hole just below it, then through the hole on the first end bar. It is pulled tight before it is fastened to the second nail. It is important that this wire is taut; its purpose is to prevent the wax from sagging. I use a pair of needle-nose pliers, bracing them against the ear of the top bar to gain purchase as I tighten the wire before wrapping it securely around the second ½-inch nail. I hold it in place with one hand, while driving in the nail and then breaking the wire with the other.

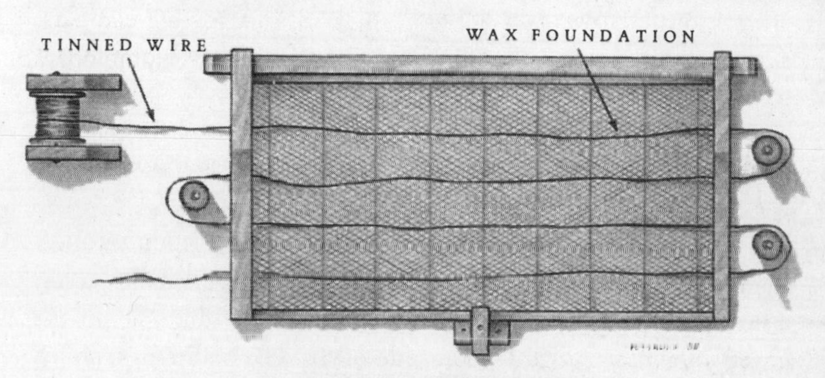

Frame with wax foundation and wire in place on frame rest

The wire is now in place, but it must be melted into the wax foundation before it can reinforce the sheet of wax. To do this, I use an electric wire embedder run from a transformer, which sends a current of electricity through the wire strong enough to melt the wax around it. For a few frames, a simpler, less expensive "spur embedder" can be used. It is a hand-held tool with a small wheel at the bottom. It is kept in a pan of hot water, which will heat the wheel enough to slightly melt and embed the wax as it is rolled across the wire.

When all these pieces of wood, wax and wire have been put together, one has a beehive.

Assembling fragrant new pine beehive parts and working with delicately scented creamy white beeswax in a snug barn made warm by a fire in the woodstove is an altogether superior way to spend a cold snowy day. It is so much fun that it is stretching matters to call it work.

In between storms, winter in the Ozarks is often clear, bright and sunny. Sparkling fair weather like that makes me think up outside work to do—work that didn't seem strictly necessary the day before, during a snowstorm. On those days, I'll sometimes pack up a lunch, fix a thermos of coffee and drive around to my beeyards to see how the bees are wintering. It is good to assure myself that all the telescoping covers are tightly in place, and to check whether cows have knocked over any hives. But the truth is I just miss the bees, and I want to see them.

The snow has drifted up against the hives, and I stoop to brush it away from the entrances. In one yard, a tree limb has fallen across a hive, knocking the cover askew. I haul off the limb, adjust the cover. In another yard, I find a pile of dry bee bodies in front of a hive and look in the snow for a clue. There I find splayed openhanded prints, complete with the indentations from a "thumb." An opossum has been here. He has reached into the entrance of the hive with that clever paw of his, stirred up the sluggish bees, and, after they have flown out one by one to defend themselves, he has caught them and sucked them dry of honey and soft body parts, leaving behind the dry husks, which look like the debris left over after a human shrimp feast.

By midday, I am at one of my favorite beeyards. The bees in this yard usually make more honey than at any other. There is not much left of the farm on which it is located. The people who own the land have retired and moved to town. Their house was torn down several years ago, and the pasture has been rented out for hay. No one has fertilized the fields for a long time, and wild things are reclaiming them. There are blackberries and sumac all around, but I have other beeyards where they grow as thickly, and the bees there do not produce as much honey as they do here. I don't know why this is such a good beeyard. The only time I ever believed that I knew all there was to know about beekeeping was the first year I was keeping them. Every year since I've known less and less and have accepted the humbling truth that bees know more about making honey than I do.

This is a pretty beeyard, too. The hives sit in a grove of pine trees. There is a mulberry tree behind them, with a blackberry thicket growing up underneath it. The land slopes away gently to the south, and from the knoll where the beehives are I can see for miles across a stretch of woods and overgrown fields. I like to stop here and eat my lunch—in the summertime I can make a dessert of ripe blackberries or mulberries. The bees have pollinated them well. My fruit today, though, will have to be a store-bought apple. I scrape away the snow at the foot of one of the pine trees and sit down, leaning against it so that the sun is on my face, and open my lunch.

The line of beehives is to my left, and they, too, are facing south. The sun is warming the front of each hive, tempting the bees out for the euphemistically termed "cleansing flights." Bees are exceptionally tidy little animals and will not defecate inside their hives. On a winter day warm enough for them to fly out, the snow in front of their hives is spattered with yellow droppings which have been retained inside their bodies. If they do not have a flight day every few weeks, they will sicken rather than eliminate their metabolic wastes within their hive.

The textbooks say bees cannot fly unless it is 10° C. or more. The bees have not read the textbooks and often fly out for their cleansing flights on days like today. The snow is melting only in protected spots where the sun warms it, so the air temperature can't be much above freezing, but the bees in this sheltered pine grove are able to fly. They won't be able to stay out long, however, away from the warmth of their cluster. One lights on my bare hand and walks along it. I open my palm, transfer her from one hand to another. I can feel each of her six legs as she walks across my skin.

Last summer, some friends came over for a visit. I had been restacking bee equipment, and there were bees flying around drawn by the odor of the honeycomb. Although the bees were not paying attention to them, my friends were apprehensive and retreated behind the doors of the cabin. Their six-year-old daughter chose to stay with me while I finished putting the equipment away. She was wary, but not yet as fearful as her parents, so I put a drop of honey on my arm to draw a bee. In a few minutes one had landed and extended her proboscis. I showed the girl the bee's long grooved tongue, which she was using to suck up the honey. The child was delighted and demanded her own drop of honey. I obliged, and soon we each had a bee feeding on our arms. When the drop of honey was gone, the bee began to investigate the rest of the girl's arm, carefully picking her way among the hairs growing on it. The child began to giggle.

"I can feel its little feet," she said. "They tickle, but I like it."

She wanted a drop of honey on her other arm and soon another bee found that. We lured several more, and after they had all flown away she ran laughing into my cabin to tell her parents what it felt like to have bees walk on her skin. I remember that child as the bee walks across my hands and think she would enjoy sitting here with me today. At last, my bee decides it is time to return to her sisters, and she flies back to her hive, the third one from where I am sitting.

In the comparative midday warmth the bees are attending to

other sanitation chores. Winter bees live several months longer than summertime ones, who may wear their wings to nubbins and die in six weeks, but over the course of the winter there are always a certain number who die. On a day warm enough to tend to them, the worker bees carry out their dead sisters and drop the bodies some distance from the hive. It is a sign of a healthy hive that the colony is strong enough to take care of this necessary task. Leaning away from the pine tree, I can see the front of every hive in the row, and all along it there are workers carrying out their few dead, each struggling with a weight nearly as great as her own.

Bees evolved along with the flowering plants back in Cretaceous times, probably somewhere on the continent we now call Africa. From their tropical homelands they spread gradually into temperate regions. When human beings came along, they learned to appreciate bees, and have been allied with them at least as far back as there are records (a rock painting in the Cuevas de la Arena near Bicorp in Valencia, Spain, dating from 7000 B.C., shows a man gathering honey from bees) and have helped them spread to places far from the tropics. But bees are conservative creatures, and maintain a steady tropical warmth within their home no matter what the outside temperature may be. In wintertime, they must generate heat to keep it the way they like it. Even during the early winter, as the light wanes each day and the queen lays no eggs and the bees raise no young, they will not let the temperature sink much below 20° C. within the cluster. In January, after the queen has begun to lay eggs, the bees raise the temperature an additional ten degrees or more, to keep their young brood warm.

Bees generate heat within the cluster by simply metabolizing fiercely. Honey is a dense, heat-producing carbohydrate food. One tablespoon of it contains sixty-five calories. And bees are efficient converters of those calories into heat, which they preserve by forming a tight insulating cluster to hold the warmth. One researcher measured bees' heat-generating capabilities and found that at 10° C.

a single bee can produce at least one-tenth of a calorie of heat per minute. Presumably a colony of even small wintertime size of twenty thousand to forty thousand bees is capable of producing thousands of calories per minute.

Depending on the outside temperature and the number of bees

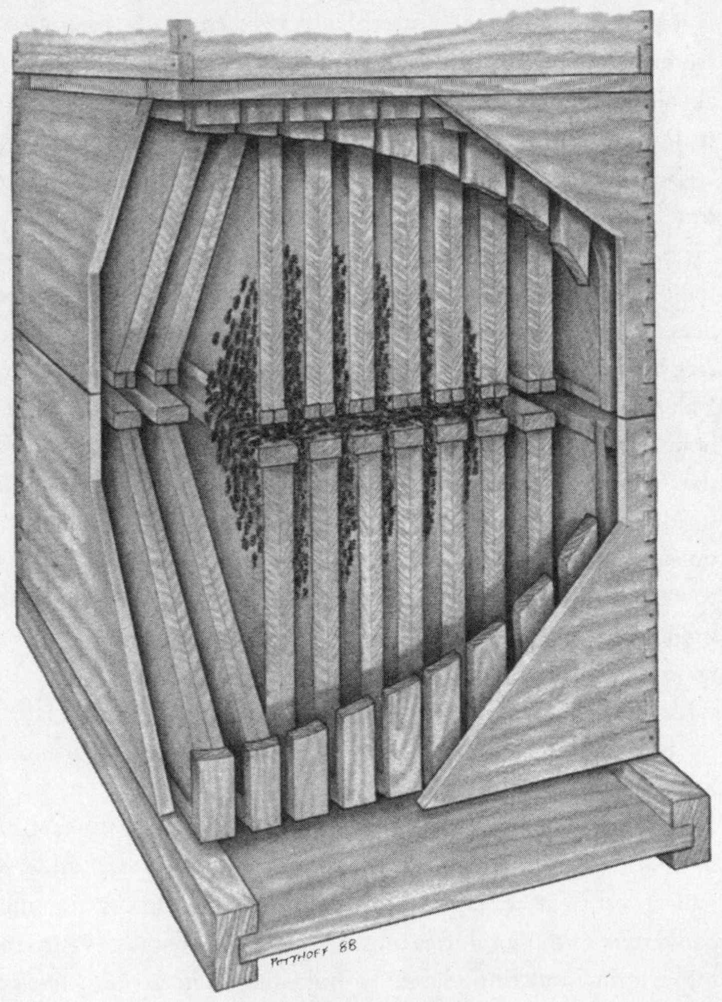

Bees in a cluster in their hive

in the colony, the cluster may extend across several frames, with bees linking one part of it to another across the end bars.

During very cold weather, the colony of bees clusters tightly; on warmer days, the cluster expands slightly. On mild days, when melted snow runs in rivulets down the gravel roads, the cluster breaks up completely and the bees fly out of their hives. In protected places along the south face of creek hollows, they may discover the pollen-bearing catkins of hazelnut or even unseasonably early maple blossoms. But significant sources of new food—serviceberry blossoms and the general bloom of maples—are yet months away, and if the colonies are to survive they must rely, at least until March, on honey made from last autumn's asters. As bees eat up the honey stores in one part of the hive, the cluster gradually moves to another part.

Winter feeding of bees whose stores have been depleted is a risky procedure and may do more harm than good. Chilled, they will drown in the liquid honey or sugar syrup in a feeder. It is possible to give them a frame of honey from another hive, which they can use directly, but in order to do so the donor hive, too, must be opened and the seals that the bees have so carefully made against the cold must be broken. This stresses both colonies. Any bees that need to be fed should be fed early in the autumn, and even then I do it only on an emergency basis. If a hive is too low in stores to make it through the winter, I prefer to combine it with an abler hive in the autumn, because even if it makes it through the cold weather it will not be strong enough to prosper in the springtime.

The low sun is no longer shining against my pine tree, so I move to another. I can see the fronts of the beehives better from here, too. I pour myself a cup of coffee from the thermos and settle down to watch the bees attending to their wintertime chores.

In more northerly parts of this country and in Canada, where winters are long and severe, many commercial beekeepers gas and kill their bees at the end of each honey harvest. They can then take all the honey the bees have produced during the long daylight hours of a northern summer. In the spring, they start anew with package bees shipped from southern bee breeders. They consider this the only economical way to produce honey. A few years ago our state beekeeping association was addressed by one such commercial operator from Nebraska, who laid out the savings in labor costs in not overwintering the bees and totaled the cash benefits from being able to harvest a larger honey crop. The audience was quiet and attentive, and I thought perhaps I was the only one who found his methods repellent and unacceptable. I was surprised and pleased, however, when one of the most prominent commercial beekeepers in our state, a man with thousands of hives, jumped to his feet. He is a tough ex-marine, a shrewd businessman and no sentimentalist, but he passionately condemned the Nebraskan's methods as cruel. He put into words what most of us at the meeting, as it turned out, believed: that was an ungrateful way to keep bees. We applauded him for a long time.

I thought of the meeting while I sat watching my bees. It makes no business sense for me to spend the time and money to drive around to my beeyards in midwinter. I might better have spent my day in other ways than leaning against a pine tree in the pale sunshine sipping rapidly cooling coffee. I am not helping the bees now. They do not need me. But I need them.

I bought my first beehives in the autumn, and then spent the following winter trying to figure out what to do with them. I read everything I could find in print about bees. I talked to as many beekeepers as would stand still for my questions. I joined the state beekeeping association and heard talks by the experts at its meetings. I found there are as many ways to keep bees as there are people to tell about it.

Beekeepers are an opinionated lot, each sure his methods and his methods alone are the proper ones. When I first began with bees the great diversity of passionately held opinion bewildered me, but now that I have kept bees in widely scattered locations I think I understand. My bees cover one thousand square miles of southern Missouri in their foraging flights. I have detailed climatic maps of that rugged area, and they show that some beeyards, only a few miles apart, are in different climate zones. Frosts come earlier in some places than in others. Spring comes later. Rainfall is not the same. The soils and the flowering plants they support are unlike. Through the years I have learned that, as a result of all those variations, I must keep the bees variously. Most people who keep bees have only a few hives and keep them all in one place. They find it difficult to understand how practices they have found successful do not work for others. But I have learned I must treat the bees in one yard quite differently from those even thirty miles away. So it is no wonder that what works well with bees for a writer based in Vermont may not work at all for one based in Arizona. The thing to do, I discovered, was to learn from the bees themselves. After a person learns something of bee biology and behavior, he can make up his own rules—and then can have the fun of defending them passionately to other beekeepers.

Beekeeping meetings are lively affairs, with hot arguments raging over the best methods of requeening, how to administer antibiotics and the proper way to super. Books and articles written about beekeeping are similarly warm and strong.

I have never counted entries in a major research library catalog, but I suspect more has been written about bees than any other animal with which we share this planet.

Beekeeping is farming for intellectuals. The Greeks spun tales about the god of beekeeping, Aristaeus. Pliny wrote about bees. Aristotle observed them, puzzled over them and reported his findings. Virgil made bees the subject of his fourth Georgic, a part of the series of poems with agricultural themes. Classicists insist that Virgil's purposes were political, that he used bees and other agricultural motifs as a vehicle, yet it is clear that Virgil knew bees and loved them. His fourth Georgic is a beautiful poem, one that gives me goose bumps when I reread it sitting beside the woodstove fire on a winter evening. Quoting from L. A. S. Jermyn's translation, it begins:

> Honey that's borne upon the winds of heaven,
> A gift of the high gods, is now my tale.
> On this page too, Maecenas, turn thy glance.
> For look! amazement grows as thou beholdest
> Great-hearted leaders in such tiny states,
> Orderly custom, yes! and national aims
> Tribes and their tribal conflicts. Small the theme,
> But yet not small the glory, if such gods
> As frown thereon, permit the tale in song,
> And if Apollo hearken to my prayer.
>
> In the first place, we must seek out for bees
> A habitation free from wind that checks
> The airy robbers winging booty home,
> A place where sheep and butting goats tread not
> The flowers, nor hooves of wandering heifer sweep
> The dew from off the plain, nor crush the plants
> That newly spring. Far from the brimming cells
> Be lizards driven—for all their painted scales!—
> Bee-eaters too, nay, birds of every kind,
> And Procne in particular, her breast
> With blood-stained claws imbrued; for these make vain
> Your labour everywhere, and carry off
> In snapping bills the flying bees themselves,
> Sweet-savoured morsels for young beaks agape.
> But let clear founts be nigh, or moss-grown pools,
> Or tiny rivulet hurrying through the grass:
> Let palm or huge wild olive shade the porch,
> That when the young kings lead the early swarms,

And, issuing from the hive, young warriors sport
In the springtime which is their very own,
A bank nearby may woo them from the heat,
And fronting trees with hospitable shade
Enfold them . . .

Columella, the Roman writer on agriculture, knew his bees, and although nearly twenty centuries have passed since he wrote, his elegantly phrased advice still bears consideration:

> If thou wilt have the favor of thy bees, that they sting thee not, thou must avoid such things as offend them: thou must not be unchaste or uncleanly; for impurity and sluttiness (themselves being most chaste and neat) they utterly abhor; thou must not come among them smelling of sweat, or having stinking breath, caused either through the eating of leek, onions, garlick, and the like, or by other means, the noisomness whereof is corrected by a cup of beer; thou must not come puffing or blowing unto them, neither hastily stir among them, nor resolutely defend thyself when they seem to threaten thee; but softly moving thy hand before thy face, gently put them by; and lastly, thou must be no stranger to them. In a word, thou must be chaste, cleanly, sweet, sober, quiet, and familiar, so they will love thee . . .

Throughout the centuries, bees continued to attract the attention of able and literate observers. Some of what has been written about them, because of the intellectual stature and abilities of the writer, is enduring, and deserves a place on the bookshelves of anyone interested in good writing about nature.

In 1609, Charles Butler, an Englishman who also wrote books on logic, music, English grammar and the marriage of cousins, published *The Feminine Monarchie, Written out of Experience,* in which he daringly asserted that the King Bee was no male at all but a female and must henceforth be called Queen. In addition to being a close and accurate observer, Butler was a fine writer. The temptation is to quote page after page of his writing, but I shall hold myself to one small sample. He characterizes a drone, or a male bee, as

> a grosse Hive-Bee without a sting, which hath always been reputed a greedy lozell: for howsoever he brave it with his round velvet cap, his side downe, his full paunch and his lowd voice; yet he is but an idle companion, living by the sweat of others brows. For he worketh not at all, either at home or abroad, and yet he spendeth as much as two labourers: you shall never find his maw without a good drop of the purest nectar. In the heat of the day he flieth abroad, aloft, and about, and that with no small noise, as though he would doe some great act: but it is onely for his pleasure, and to get him a stomach, and then returns he presently to his cheere.

François Huber was a Swiss naturalist who worked with bees in the late 1700s and early 1800s. He was blind, but with the help of his wife and a servant to make observations he wrote beautifully about bees. His dramatic description of a jealous virgin queen bee destroying her potential rivals before they have emerged from pupation has never been equaled:

> Hardly had ten minutes elapsed after the young queen emerged from her cell when she began to look for sealed queen cells. She rushed furiously upon the first that she met, and by dint of hard work, made a small opening in the end. We saw her drawing with her mandibles the silk of the cocoon which covered the inside. But probably she

> did not succeed according to her wishes, for she left the lower end of the cell and went to work on the upper end where she finally made a larger opening. As soon as this was sufficiently large, she turned about to push her abdomen into it. She made several movements in different directions until she succeeded in striking her rival with the deadly sting. She then left the cell and the bees which had remained so far perfectly passive began to enlarge the gap which she had made and draw out the corpse of a queen just out of her nymphal shell. During this time, the victorious queen rushed to another queen cell and again made a large opening, but she did not introduce her abdomen into it, this second cell containing only a royal pupa, not yet formed. The young queen rushed to a third cell, but she was unable to open it. She worked languidly and seemed tired from her first efforts.

Unfortunately elegance and vivid style is no longer the fashion in entomological writing, but contemporary authors have much to tell us, and their books and articles are good reading on a long winter evening. Among the academics, J. L. Gould at Princeton and Roger Morse, from Cornell, both have done important work with bees and written of it. Entomologists associated with the federal bee laboratories under the Department of Agriculture, such as E. H. Erickson, publish the results of their research in the beekeeping magazines and entomological journals. The federal bee laboratories, scattered across the country, have been victims of budget cutting in recent years, but in the past they have generated both basic and applied research in apiculture. As a result, there are many helpful beekeeping pamphlets that may be purchased through the Government Printing Office.

There are three basic "texts" for beekeeping:

The Hive and the Honey Bee. Dadant & Sons, Hamilton, Illinois.

The ABC and XYZ of Bee Culture. A. I. Root *et al.,* A. I. Root Co., Medina, Ohio.

Honey: A Comprehensive Survey. Ed. by Eva Crane. Crane, Russak, & Co., New York.

These three books are compilations of contributions by many authors, and as a result the books vary throughout in quality and readability. The first two are published by beekeeping supply companies and therefore lack a certain critical distance from the subject covered. All three contain more than the beginning beekeeper wants to know. He shouldn't read them as an introduction to beekeeping, because they will overwhelm him, but he should buy them and save them for his second winter's reading. They will be useful reference books for him later. Nearly every beekeeping supply company publishes a short introduction to beekeeping, and there are several others written for specific parts of the country by independent authors. Richard Taylor, a metaphysician and master beekeeper, has written a good guide for the Northeast, *The How-To-Do-It Book of Beekeeping.* Here in the middle part of the country, beginners can profit from *Beekeeping in the Midwest,* by Elbert R. Jaycox, an academic entomologist, a book available from the Agricultural Publications Office of the University of Illinois at Urbana.

In addition there are many books on bees that are simply pleasant to read. One of the best is *The Dancing Bees,* by Karl von Frisch, a popular account of some of the work that earned him a Nobel prize. My own favorite is Frank R. Stockton's classic, *The Bee-Man of Orn,* a children's tale reissued in recent years with marvelous illustrations by Maurice Sendak.

The above list is eclectic. I have left out important authors and important books. But once a person starts reading anywhere in the beekeeping literature, one author introduces the reader to another. One winter's reading generates the next winter's booklist.

Aristaeus, that god of beekeeping, turned up as the villain in the celebrated and powerful film, *Black Orpheus,* the retelling of the love story of Eurydice and Orpheus set in Rio during the Carnival.

Their story has been told and retold in opera, but Aristaeus has a story of his own.

According to the ancients, Aristaeus was the son of Cyrene, a naiad who despised spinning, weaving and similar housewifely tasks. Instead, she preferred to hunt wild beasts all day and half the night. Apollo once watched her wrestle a lion to the ground and was so struck with love for her that he carried her off to Africa and built a palace for her there, in the place that still bears her name. After their son Aristaeus was born, Apollo left her, and Cyrene, yearning once again for wild places, left too, leaving her son to be raised by a group of myrtle nymphs who taught him how to curdle milk for cheese, keep bees in terra-cotta pots and cultivate olives.

When he was grown, Aristaeus left Libya and traveled among human beings, who accorded him divine honors for the useful arts he taught them. He also practiced healing, and once cured the plague by changing the direction of the winds.

In his wanderings, he met Eurydice, a wood nymph, the beloved of Orpheus. He tried to rape her, but she ran from him through the woods and in her terror did not see a large poisonous snake in her path. Tripping on a tree root, she stumbled and fell. The snake struck her and she died from the poison of the bite. Orpheus, heartbroken, went down into Tartarus to fetch her back, playing his lyre as he went. With his music he not only charmed the guards and the dog Cerberus at the gates of Tartarus, but temporarily suspended the torments of the wicked as well. Hades himself, lord of the underworld, was so moved he agreed to allow Eurydice, guided by the lyre music, to follow Orpheus back to the world of the living, provided that Orpheus did not look back at her until she was safely in the sunlight. But Orpheus so loved Eurydice he could not keep himself from looking to see whether she was following, and so he lost her forever.

The other gods were filled with wrath at Aristaeus and punished him by killing all of his bees. He had no idea why they had died, so he set out on a journey to find his mother, mistress of wild things, and ask her for help. After a long time, he found her living under a stream with other naiads. She knew nothing of bees but suggested that he go see her cousin, Proteus, he of the many shapes, and ask him why the bees had died. Aristaeus found Proteus and wrestled him to the ground to force him to hold one shape until he revealed that the bees had been destroyed by the gods.

Armed with that piece of information, Aristaeus went back to Cyrene and asked her advice. She and the other naiads consulted and determined that he should return to his home, where he should sacrifice four bulls and four heifers and pour blood over them. Then he should go away, leaving the carcasses, and return nine days later, in the morning, with poppies of forgetfulness, a fatted calf and a black ewe to propitiate the ghost of Orpheus, who by this time had joined Eurydice's shade in Tartarus.

Aristaeus did as he was told, and when he returned on the ninth day he found a swarm of bees emerging from each of the carcasses. He captured them and put them in his empty terra-cotta beehives. Aristaeus was grateful for the gods' forgiveness and settled with his bees in Boeotia, where he married Autonoë. To them were born two children: Macris, who was to become nurse to Dionysus, and the ill-fated Actaeon. After Actaeon had grown to manhood, he was out hunting one day with his dogs and quite by accident came upon a pool in which the goddess Artemis was bathing. He was so struck with her naked beauty that he stood and watched. Artemis was enraged when she discovered him, and punished him by changing him into a stag, whereupon he was torn to bits by his own pack of fifty hounds.

Aristaeus was grieved by his son's death and left Boeotia to begin traveling again. He went back to Libya, to Sardinia and distant islands, spreading the knowledge of beekeeping and the agricultural arts. Near Mount Haemus, in what is now known as the Balkan range, he founded the city of Aristaeum, and then, as a god may do, disappeared without a trace.

III

THE BEEKEEPER'S SPRING

The Beekeeper's Spring

Spring for bees, and so for beekeepers as well, has nothing to do with the calendar. In protected hollows even in the Missouri winter, bees can find a bit of springtime in flowers with enough pollen to feed the first young bees growing from the eggs the queen has begun to lay.

Various species of maples, all members of the genus *Acer,* with their delicate blossoms ranging from yellow-green to red, are one of the earliest to bloom. Maples are tall trees, and their blossoms are borne high in the crowns, where they often are not seen by humans. The bees, of course, have no trouble in finding them. The American hazel, *Corylus americana,* is a smaller tree—hardly more than a shrub—that is another early source of pollen, and easier for us to see. It grows wild throughout most of temperate North America and produces the nuts we call filberts. In late winter and early spring when most other plants are at rest, hazel generously produces pollen on long, dangling catkins. In the winter, I like to stop on my walk down to the mailbox and draw one of the yellowish brown catkins between my thumb and forefinger and watch the powdery golden pollen collect in my palm. On days warm enough, bees, having declared it springtime, will be on the catkins collecting pollen.

Skunk cabbage, *Symplocarpus foetidus,* with its coarsely sheathed flower and supposedly vile odor, makes spring for bees in February in northern parts of the country. If a walker in the wet woods or swamps takes the time to look inside the skunk-cabbage spathe on a mild day, he will usually find a honeybee there, packing pollen into the carrying baskets on her back legs.

Mature adult bees are carbohydrate feeders. They live on nectar and honey, which are almost exclusively carbohydrate in composition except for a few traces of minerals, vitamins and other materials. But young and developing bees need protein to grow muscles, glands and other tissues, and it is pollen—variously rich in protein, depending on its source—that provides it.

It takes about three weeks for a bee to grow from egg to adult. Three days after the queen bee has laid an egg, it hatches out into a nearly white larva that, throughout its instars, or developmental stages, has a ferocious need for food high in protein. Bee larvae don't move around looking for food as some insects do in their early stages; they are not wriggly caterpillars as are those of moths or butterflies. They merely lie passively in their cells, growing, changing. They are small animals, with disproportionately large stomachs waiting to be filled by nurse bees—young adults just emerged from pupation. These nurse bees feed on the gathered pollen and produce secretions from their hypopharyngeal glands (glands under the pharynx), which they will use as food for the developing larvae.

In recent years, vendors of nostrums to the health-food industry have been delighted by Ronald Reagan's public consumption of pollen, and pollen has become a hot item in health-food stores. Claims are made that humans who eat pollen are stronger, sexier, livelier and more cheerful, but there is no good medical evidence to back up these claims so far. And as a matter of fact, some people have experienced severe allergic reactions after eating it. Pollen from a hay-fever sufferer's local area is also said to cure him of his allergy. I have an open mind. I am an empiricist. I have hay fever. And, after all, bee venom does help my arthritis. So one summer, when my eyes were running and I was sneezing exquisitely from ragweed, I gathered and ate pollen from my bees. Pollen has a bitter taste. It is no fun to eat, but in the interests of testing the efficacy of the stuff I dutifully continued to eat it—and to sneeze. It is possible pollen may cure someone else's hay fever, but it doesn't cure mine.

I am always too eager to start working with the bees in the spring. I miss them in the winter and want to be back with them again. But I sternly remind myself I can do more harm than good if I open their hives before the daffodils bloom, so when the crocuses flower around my cabin I have to content myself with crouching down beside those purple, white and yellow blossoms to watch the field bees gathering pollen.

Each worker bee is covered with stiff hairs. These catch the granules when the bee brushes against the flower's pollen-bearing anthers. She is soon covered and stops to use her two front pairs of legs to comb the sticky pollen from her body and pack it into the baskets on her rear pair of legs. When her legs are heavy with their wads of pollen she returns to her hive to unload. House bees arrange the pollen granules in the honeycomb cells around the developing bee larvae, where it is conveniently available to the nurse bees.

In the springtime, a typical frame of brood—as beekeepers call developing bees—contains a large central semicircle of cells containing eggs and larvae in various stages. The eggs and larvae are there to see, but the oldest brood, stomachs full, have spun themselves cocoons in which to pupate, and are hidden from human eyes, sealed in their cells by a tough, velvety brown covering, from beneath which they will emerge as adults. The entire semicircle of brood is bordered by a ring of pollen-filled cells, which in turn is surrounded by honey, or, as the season progresses, nectar.

In this part of the country most of the early pollen collected is yellow and makes a bright ring around the fuzzy brown sealed brood area. But as the season advances and the days lengthen, more pollen sources are available. Serviceberry, *Amelanchier arborea,* frosts the slopes of the hills near the rivers with white blossoms. Redbud, *Cercis canadensis,* flourishes in the wood's understory, with dramatic purplish red clusters of blossom. Henbit, *Lamium amplexicaule,* which everyone except beekeepers think of as an unwelcome weed, covers the ground with a low-growing thatch of purple-blue flowers. The wild fruit trees—plum, cherry and peach—bloom extravagantly. These flowers all have different-colored pollens, and the pollen cells ringing the bees' brood are as various: scarlet, pale green, yellow and orange—as beautiful as a stained-glass window.

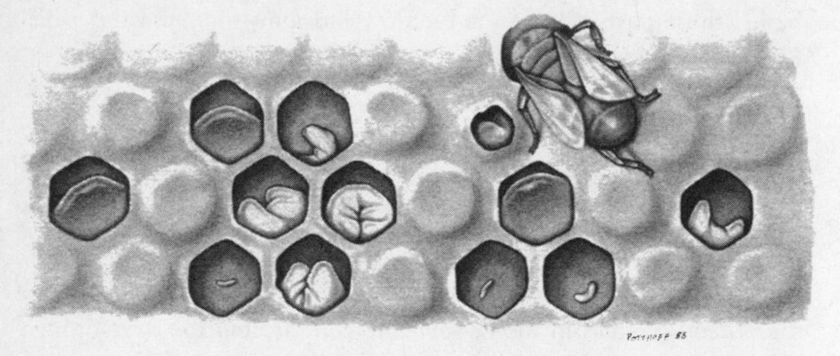

Typical brood frame with sealed brood in center, surrounded by uncapped brood, pollen and honey

In early springtime, the bees' need for pollen may be greater than what is available, and then they will compulsively gather up almost any fine granular material—sawdust, bits of plastic packing material, ashes—whether it is useful to them or not. I often see them at the bird feeder, picking through the wild bird feed. At that time of year, I often get a telephone call from a dairy farmer who lives near one of my beeyards.

"Hey, Sue," he says. "How about coming over and feeding your bees? They're in the calf feed again."

He's always a little embarrassed. The first time he called me, years ago, was on a day in autumn. There were, he insisted, bees flying all around the inside of his milking shed. He didn't want to spray them with insecticide, but he was going to have to do that unless I could lure them out.

It sounded like an odd way for bees to behave, but I drove over to his place. There they were, golden-striped fuzzy insects flying around inside his barn, darting here and there. They looked like bees, but they didn't act like bees, so I captured one in my hand. The creature did not sting me as a honeybee would have done and on close scrutiny its beelike appearance disappeared. Moreover, the insect had only a single pair of wings, not two sets, as a bee has. The two wings, a single pair, puts any insect into the zoological order Diptera (which means two-winged), the order of flies. Honeybees belong to the order Hymenoptera (or membrane-winged), which includes not only all the bees but also wasps, ants and other similar insects.

I wasn't sure which flies these were in the milk barn, because I don't know much about Diptera, but I knew that there were a number of fly species that mimic bees in appearance. I showed the farmer the single pair of wings and told him that this and the others in the air were flies, not bees. He was interested, and asked more questions than I could answer. When I got home, I photocopied as much information about bee-mimicking flies as I could find on my bookshelves and sent it to him. Now when he calls he always makes sure that the insect about which he is lodging a complaint is indeed a honeybee.

The first time he found bees in his calf feed he was worried that they were after the calves, and might sting and frighten them. I assured him that the bees were not the least concerned with the calves but were looking for the pollen they needed as food in the springtime. As soon as some pollen-bearing flower bloomed, I told him, the bees would disappear from the calf feed, but in the meantime I'd try giving them a pollen substitute that might satisfy them.

Beekeeping-supply companies sell a commercial pollen substitute made of a combination of protein-rich materials such as soy flour and brewers' yeast. This powder can be mixed with honey to make into patties, which can be squished down between the top bars on the frames of the upper hive bodies. Many beekeepers recommend the feeding of pollen substitute as a regular routine, and for a number of years I made up patties and fed them to all my hives. In some parts of the country where there is not a dependable source of pollen this may be a good practice, but I have concluded that it is not worth the expense or time here in the Ozarks. Bees prefer the real thing—fresh pollen—and although a day or two in these hills may pass when they can't find any, something usually bursts into bloom and they scorn the substitute, leaving it to molder inside the hives. So I stopped feeding pollen substitute, and find that productivity of my hives is no poorer for doing so. Indeed, the per-hive yield of honey from my bees has grown greater in what appears to be a direct relationship to the decrease in the number of times I open the hives. The less I disrupt and fiddle with the bees, the more they can concentrate on making honey.

Now when the dairy farmer calls, more as a public-relations gesture then anything else, I go to the health-food store and buy fifty-nine cents' worth of soy flour, take it over to the hives near his place, lift off the outer cover of each hive and sprinkle some of it on the inner cover. I did that a few days ago, and today I stopped by his place to give him a jar of honey as a recompense for the nuisance my bees had caused him.

I asked if the bees had stopped coming to the calf feed.

"Yep. They've left. Thank you kindly for feeding them."

"No trouble. Thank you for calling me and letting me know. I don't like the bees to be a bother to you."

Then I drove over to the hives and lifted up the telescoping covers. The soy flour stood unused on the inner covers, and I shook it off on the grass. The bees had not wanted it. Flowers have begun to bloom somewhere, and the bees, their rear legs wadded with golden pollen, are flying into the hives so heavily loaded they find it hard to stay airborne.

The farmer is happy. The calves are happy. The bees are happy.

I sit down beside the end hive and watch the bees flying in with their loads of pollen, which will assure the development of the thousands of young bees needed to gather nectar and make honey in the months to come.

The sun is shining on my back as I watch, and I can feel its warmth. The air is fragrant. There are flowers in bloom everywhere today.

And I am happy.

The daffodils finally do bloom. They do every year, of course, although in some years when I want to get out among the bees it seems as though they never will. The golden blossoms surround my cabin, but they also march off in straight lines through the underbrush and scrubby sumac, wild cherry and farkleberry which are the first advance growth of the woods that is closing in around my place. The widow from whom I bought this farm fifteen years ago laughed when she told me about those straight rows of daffodils. She had bought a a burlap sack of bulbs, and had given them to her husband to plant.

"He tried digging holes for 'em by hand," she said, "but after a few he decided he weren't going to bother that hard, so he just hitched the cultivator to the mule and make him some long straight furrows and dropped in them daffodils. I thought they looked pretty funny, but I never did say anything, and now they make me think of him and his mule and that day and wish it were all now."

When I see those daffodils striding into the young woods, I like to try to create in my mind that day of theirs, too, and remember the good things they did to establish this farm where I have had bees and been happy. I've come to the belief that we manufacture whatever immortal souls we have out of the bits of difference we make by living in this world. It seems no bad thing to have a soul of yellow daffodils in lines across an Ozark hilltop.

The next several weeks after daffodils bloom are the busiest ones for a beekeeper. I'll spend every day working with the bees from early morning until the sun sets. If I had just a few hives, I could save the work for the middle part of sunshiny warm days. Bees are easiest to work with then, because most of the older bees are out foraging. The ones who remain are young house bees, less likely to take offense at having their hives opened, less likely to sting. But because I have many hives to work through, I must risk irritating the bees and take some stings. I'll be opening the hives early and late and on rainy days as well as fair ones. During the weeks to come, I'll visit every one of my beehives at least twice. I'll be checking to see that each hive is strong and healthy and I'll medicate them to prevent disease. I'll be feeding those I must and preparing some of the strongest hives for division into two. I'll be cleaning up any mouse damage I find and replacing worn beehive parts and combs that have become distorted and dark from use over the years. I'll be doing everything I can to make sure I have strong populous beehives by the time the wild blackberries—which include dewberries and several other species of the genus *Rubus*—bloom. Those blossoms produce the first major nectar flow, and from it the bees will make honey that I will be able to take from them at harvest time. Before the blackberries begin to bloom, the bees must use up all the scattered nectar-producing sources in order to keep the colony going and to build up its strength. In other parts of the country the first strong nectar flow comes from other flowers; a beekeeper must be something of a botanist in order to learn the blossoming patterns of flowers that represent significant nectar flows for the bees.

This is one of the most enjoyable aspects of beekeeping, but it took me several seasons to convince myself that I was doing genuine productive work when I walked out into wild places with field guides to the wildflowers and shrubs, a notebook and a picnic lunch (and, if I could persuade a friend to come along, a bottle of wine) in a knapsack. It seemed like too much fun.

The day before I go out for my first trip to the beeyards I prepare the medicine I'll need for each hive and the sugar syrup I may need for some.

Of the number of bee diseases that can have a debilitating effect on a colony, none are as serious as American Foulbrood. It can be prevented by medicating bees beforehand; it cannot be cured once they have it. Bacteria are ubiquitous and occur in immense number and variety. Most of them are helpful and benign. Few cause disease but one, *Bacillus larvae* is the agent responsible for American Foulbrood in bee larvae.

The bee larvae become infected by eating food contaminated with the spores of the bacteria, which is so persistent that it can remain infective for thirty-five years. The only remedy for a hive with American Foulbrood is to burn it—bees, honeycomb frames, hive parts and all—for after the bees have died out in a sick colony, as they will almost surely do, any healthy bees put into the hive will become infected themselves.

From time to time, bee breeders have worked to develop strains of bees resistant to American Foulbrood. Right now, Steve Taber, a West Coast entomologist of great skill and wisdom, is working on this particular problem, but so far a beekeeper's best defense is to medicate all of his hives with Terramycin as a preventative. I have medicated my hives each year and have never had a case of American

Foulbrood in them, although it is common in this area. The bees belonging to beekeepers who fail to medicate, and wild bees in trees, are potential reservoirs of infection. When any colony of bees, wild or hived, begins to sicken, it becomes too weak to defend itself from robber bees, and foraging workers from healthy hives steal honey from a hive ill with American Foulbrood and carry the bacteria back to their own brood.

Affected brood die after they have spun cocoons and are sealed in their cells. They turn brown, putrefy and give off a fetid odor. The cappings over such larvae become dark and moist, sink inward and are sometimes marked by holes made by adult bees that try to clean out the infected cells. The test for suspicious-looking brood is to dip a matchstick or twig into the putrefying larva within a cell; if the fluid residue stretches out into a ropy thread, the chances are that the bees have American Foulbrood. Sure identification, however, is possible only from laboratory analysis. Most states have inspectors who can help with identification of bee diseases, and the federal Bee Laboratory in Beltsville, Maryland, can provide positive identification from fragments of diseased comb.

There are a number of other bee diseases that can have a debilitating effect on a colony, even though they are not as serious as American Foulbrood. European Foulbrood is similar to American in some of its symptoms, but can usually be cured by reducing stress on a colony and requeening. Terramycin also helps to prevent it. Bees succumb to other bacterial diseases as well as to those caused by viruses, protozoa and fungi. Parasitic bee mites have, in recent years, spread to North America, which had been free of them before. They are harmful to bee colonies and researchers are investigating medicines and methods to keep them in check. One of the most interesting is work on the development of a mite-proof strain of honeybees, possibly by using the Africanized bee, which, it has been discovered, has an inborn resistance to at least one of the mites, *Varroa jacobsoni.* (The bug that headline writers like to call the killer bee and entomologists call the Africanized honeybee belongs to the same species, but represents a different strain, one that is moving northward from South America, where it was imported from Africa. In many ways these bees are the same as their gentler cousins, but they are quicker to anger and will follow for a greater distance someone who has disturbed their hives. My beekeeping magazines are full of conflicting reports about them. At this time, there is no real agreement on how they will behave if and when they become established in temperate latitudes.)

The Beltsville Bee Laboratory publishes a number of helpful pamphlets on bee diseases, but the most comprehensive source I know is Leslie Bailey's *Honey Bee Pathology,* published by Academic Press. It is worth reading for a general understanding of the subject, because a beekeeper should be able to recognize a diseased colony and treat it in an informed way, but it should also be read with a sense of balance. Young medical students often decide they have each fatal condition they learn to diagnose, and new beekeepers sometimes fall into a similar habit, believing their bees have every disease they read about. Most strong colonies of bees never sicken. Most well-kept bees are healthy.

Before heading out to my beeyards I mix up Terramycin with powdered sugar. The ratio of mix depends upon the strength of the Terramycin, and manufacturer's suggestions should be followed. It should be fed in dry form, not mixed with sugar syrup, because it rapidly degrades in the presence of moisture. I put my powdered sugar-Terramycin mix in a big metal saltshaker and shake about a tablespoon of it across the top edges of the frames in each hive, being careful to avoid the center of the frame tops, where it could get directly onto the brood and kill it. The bees take it up rapidly. In ten days to two weeks, I repeat the procedure, so that a trace of Terramycin is widely spread throughout the bodies of the bees in the colony. It is important to medicate bees very early in the season, weeks before the first honey supers are put on the hives, because no

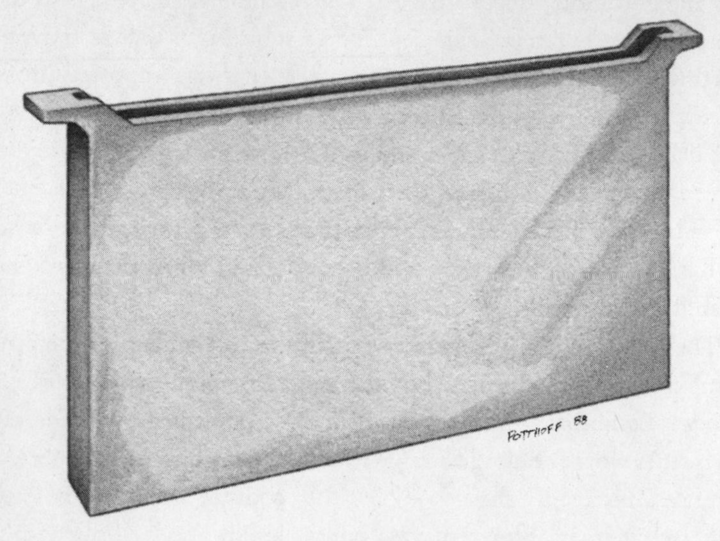

Feeder

trace of Terramycin should remain inside the hive, where it might contaminate honey to be harvested for human consumption. Some beekeepers recommend a twice-yearly dosing with Terramycin and so also feed it out during the fall. In general and on principle I do not like the routine and zealous use of antibiotics for humans or for other animals. In particular, I do not like creating the conditions in which resistant forms of bacteria are encouraged to grow, so I keep my use of Terramycin down to this one spring dosage. If, however, I hear of a beekeeper who has American Foulbrood in his hives within a few miles of mine, I will give my hives an additional treatment.

The sugar syrup I take with me on these spring bee rounds is a thin mixture made of five pounds of granulated white sugar for each gallon of water. I mix the sugar into hot water in a pail, stir until it dissolves and then pour it into five-gallon jugs that I load onto the pickup. In each bottom hive body I have a plastic trough feeder, which is just the width of regular honeycomb. If any of my hives need feeding, I can pour the syrup directly into it. It holds nearly a gallon of syrup, and that amount will keep a colony from starving to death for several days. These feeders are available from beekeeping-supply companies, which also sell several other kinds—ones that fit into the front entrances or on the tops of hives. I find the trough feeders work the best for my purposes, but using them requires taking an extra precaution for the bees' safety: until they have built a crisscross of honeycomb in the feeders, which they will do in a few years of use, I always place bits of twig inside them. The twigs float to the surface of the syrup when the trough is filled and give the bees something to stand on while they are taking it up. If they do not have a dry footing, they can drown in great numbers in the syrup—particularly on a chilly day, when they are less active. It is sad to open a beehive and find masses of bloated dead bees floating in the syrup that was supposed to sustain them.

I feed sugar syrup only to hives that are completely out of honey from the previous season and have no fresh nectar. If I don't feed the bees, they will begin to starve, burying their heads in the cells as they search futilely for food. I feed the bees to keep them from dying. Some beekeepers recommend continuous feeding of all hives in the springtime, arguing that a steady supply of sugar syrup tricks the colony into believing there is a nectar flow in progress, which stimulates the queen to lay a greater number of eggs than she normally would. I have tried it but find it unsatisfactory, and have concluded that it is impossible to trick bees. They know their world better than I do. The cost of sugar and labor involved over several months does not make good business sense in these days of low honey prices, but, more important, I have learned from the records I keep on each hive that those hives regularly fed sugar syrup end up producing less honey by summer's end than those which were not. There are a number of perfectly sound biological reasons for this curious outcome, but I like best the way Adrian Bell, the designer of the first London *Times* crossword puzzle, expresses it. He writes in his 1932 book, *The Cherry Tree*:

> Our postman said . . . "Isle of Wight disease? Never heard of it. My bees? No I never lost none. John Preach's? Why, of course they died; he used to feed 'em on syrup and faked up stuff all winter. . . . You can't do just as you like with bees. They be wonderfully chancy things; you can't ever get to the bottom of they."

Early the next morning, after I have made up the fire in the woodstove, I step outside my cabin while the coffee is percolating to get a feel for the day in the darkness before dawn. There are no clouds in the sky, and the last stars are shining brightly. The day promises fair. It is too cool to have my coffee outside and watch the sunrise, so I sit near the woodstove, where I can see the eastern sky.

The glow on the horizon reminds me that the sun will soon be up, and I have chores to attend to. The dogs, cat and chickens must be fed and watered. I have two cases of honey to pack for shipment to Dallas. I decide I'll head down to my southern beeyards first, for although they are only thirty miles away, spring will have come earlier to them. They are in a different climate zone, and flowers there bloom two weeks before they do here. I linger over my second cup of coffee, itemize in my head what equipment I need to take, plan my route. The sun rises, glowing in the morning mist. It lights up the water droplets clinging to the remains of an old spider web outside the window. The drops turn gently in the air currents, flashing blood red, green, violet, shocking pink, each droplet a spectrum of light in the low rays of the sun.

After the chores are done, I load into the pickup what I'll need: the record book, smoker, hive tool, frame grips, fuel for the smoker. I toss in some extra boards to put under hives that may need them. My record book has reminded me about several covers that need replacing, and I have put them in the truck, too. The jugs of sugar syrup are already in place. The medicine is in a coffee tin in the freezer. I take it out, make up lunch, fill a thermos with water. I am ready to go.

I light my smoker when I arrive at the first beeyard, don my gloves and bee veil and check the entrances of the row of hives. This yard is set up, as are my others, with the entrances of the hives facing southeast, so that morning sunshine hits the front of each hive and encourages the bees to fly out early and go to work. The yard is at the edge of a grove of post oak trees, which protect them from the north wind in the winter and give shade from summer's sun. Bees appreciate a bit of open shade. Those in hives set in direct sun must spend more time fanning and cooling in the summer; they have less time for flying about and making honey.

The sun is now shining brightly on the front of the hives. I can tell immediately which ones are prospering. Bees are flying out industriously from nearly all the hives but two. One of these appears to have died out over the winter. A few bees from other hives are idly investigating its entrance, and in one corner of the entrance there is a pile of wax, telling me that the colony has been plundered of its stores. There are a few bees flying in and out of the other hive, but the entrance is scattered with small dirty-white and blue flakes, which indicate that the colony within has had Chalk Brood. Chalk Brood is a disease caused by the fungus *Ascosphaera apis.* Bee larvae infected by it die, shrink and become hard. As they do so, they darken and become bluish or even black. The bees remove the dead brood and carry it away. What I see at the hive entrance are bits of brood they have not yet carried off. Perhaps there are not enough bees to clean the hive. Chalk Brood is not a serious bee disease and does not need to be treated. I seldom see it except in the spring after a period of damp weather, but the colony in an affected hive is sometimes weakened by the loss of young bees and may need some extra help.

After I have finished looking at the fronts of the hives and learned what I can, I return to the rear of the row and begin working. I smoke the entrance of the first hive, and remove its telescoping cover and inner cover. It is full of bees. There are probably more than a thousand boiling up out between the frames. They are not cross, but have simply left whatever work they were doing to investigate this intrusion into their home. Perhaps they were doing no work at all. Bees *look* terribly busy. An opened hive is the stuff of platitude, but the truth is that bees, like other animals (including humans), spend a lot of time doing nothing at all.

In an 1899 study, one Professor C. F. Hodge marked bees and watched them from daylight to dark. He reported, "No single bee that I watched ever worked more than three and one-half hours a day." In one case he saw a worker bee crawl into a cell and he watched her remain lying there quietly for nearly five hours. In the 1950s, Martin Lindauer, an entomologist, followed up on these observations. In a rigorous study, he tagged great numbers of bees and found that they spent a lot of time doing nothing at all, or very little. One typical bee, during a hundred and seventy-seven observation hours, did absolutely nothing for seventy of them and "patrolled" the nest, as though looking for something to do, for fifty-six. During roughly two-thirds of her time she was performing no productive work.

I puff some smoke across the top bars of the frames to make the bees go back down, because I want to remove a few frames and check the pattern of brood. I take out the two outside frames first, prying them loose with the hive tool and then pulling them out with the frame grips. Both still contain capped honey from last autumn. I check to make sure the queen is not on them, and lean them against the side of the hive on the ground. Then I gently pry apart the remaining seven frames to loosen and space them, so that I won't squash any bees as I pull them out. There is a scattering of honey, nectar and pollen on the outer frames. One has eggs in it. The three innermost frames are filled with brood, most of it sealed. On the very center frame, the queen walks sedately, unperturbed at being removed from the hive. I hold the frame in the grips and watch her. She is looking for an empty cell, and when she finds one she backs her abdomen into it and lays a pearly-white egg. I replace her frame carefully and return the two other brood frames around her, so that she won't be pinched and killed when I put the other frames back in place. Peering down inside the hive, I can see that the bottom hive body has only empty comb, but the bees down there are cleaning out the cells to ready it for use.

Generally speaking, the queen bee goes upward to lay her eggs, and will not fill the bottom hive body unless she is forced to. And, generally speaking, the queen is found on frames of brood. These generalities, like all others in beekeeping, are not always true. I carefully checked the inside of the top and inner cover, for instance, when I took them off, because sometimes the queen is parading across one of them; if it is placed beside the hive, she may wander off in the grass, be lost to home forever and die. This hive does have enough food to see it through, so I reassemble it, powder it with medicine and close it up.

If I had fewer hives to tend, I would do something more to it to prevent swarming some months from now. Swarming, from a beekeeper's standpoint, is not a happy event. A hive that has swarmed—or raised itself a second queen and split in two, with one group flying away—is weaker and will produce less honey. It is a perfectly ordinary procedure for bees; indeed, swarming is their way of increasing in numbers and spreading, but it is not good for their human keeper, who wishes as many bees as possible to stay in place and produce lots of honey.

Nearly every beekeeper has his own pet way to prevent swarming. The method I prefer takes into account the bees' own tendency not to develop the swarming impulse as long as they have space available inside their hive. Since the queen likes to move upward to lay her eggs, one of the best ways to give her more space is to rotate the hive bodies throughout the springtime, placing the full upper hive body on the bottom and the empty bottom hive body on the top. In a month's time, the empty hive body that had been on the bottom will be filled with brood, pollen and nectar. The brood in the other hive body will have emerged, the comb they were in will be empty and the hive bodies can be rotated again. When the hive is full of young bees, their wax glands dripping with fresh creamy wax which they urgently want to spend to build new comb, even more space can be made by pulling out several old combs that probably needed to be retired anyway and putting in their place frames that contain foundation sheets only. This gives the bees more space and it also employs the young bees in building up new comb rather than allowing them to start up that mysterious urge to swarm off to other quarters. This is not a guaranteed method of keeping bees from swarming—there are no guarantees to anything, with bees—but it does serve to keep the swarming impulse in check, in large part. It is simple, and it takes into account the bees' own biology, so it is the kind of method I prefer to the more hostile ways of preventing swarming, such as killing new queens as they pupate.

However, to be honest, I don't even follow this simple method anymore. When I had sixty, eighty, even a hundred hives, I could afford the time and labor involved in rotating hive bodies and replacing comb. But with three hundred, it became impossible. I now accept the fact that a certain number of my hives will swarm. The only thing I can still do is divide the strongest hives and provide plenty of supers at a very early date, in hopes that the additional space will keep the bees from feeling crowded.

I heft the next hive and, although it is lighter than it was in the autumn, it is heavy enough to tell me that the bees inside still have honey. They have been disturbed by the jostling of their hive, and when I open it they are in a defensive posture, their abdomens raised in the air in the sting position. I puff smoke along the tops of the frames to quiet them, and pry some frames apart to see down inside. There are more bees in this hive than in the first one, and they have five sealed frames of brood. I medicate it and mark it down in the book as one that would be possible to divide later on. After I close it up, I disturb it again by lifting the back of the hive and pushing a board under it. The boards already there have settled, and the hive has assumed a backward slope. The new board returns it to a slightly forward tilt, which is the proper position for a beehive: any moisture that collects can now run out the entrance, and the inside of the hive will stay dry and clean. The bees in this hive, already a bit on edge, are made crosser still by the disturbance, and they fly back to defend their home. I direct more smoke at their entrance and they lose interest in me.

After I have checked and medicated the next hives I come to the one I suspected had died out. I open it and find only robber bees, who are searching through the frames in a desultory way. In the center of the top hive body are four frames of starved bees, their heads plunged downward into the cells, where they desperately looked for honey to eat and, finding none, died. Below, on the bottom board, I can see a heap of dead bees who had fallen away from the winter cluster and died of the cold. The significant thing is that there was still honey in this hive when the bees died; the outermost frames in the bottom hive body contained honey very recently. Those frames have been ripped open by robber bees and their contents carried off. I can tell that because the wax cappings of the cells have been roughly torn away, the sign of the work of bees from a different colony. Bees in their own hive open honey cells neatly and carefully. This honey was too far from the winter cluster of bees, and they were never able to get to it. They must have died during a stretch of weather too cold for them to have moved—starved because they had no honey remaining in a convenient place. The bees from neighboring hives have taken the honey they left behind undefended.

I set aside the two hive bodies to take back home. Several of the bottom frames have holes chewed in them where the mouse who scampered out as I opened the hive has made her nest. She had dragged in bits of fluffy thistle head and leaves to cushion the hole she chewed in the wax. I scrape out the nest and dump the pile of dead bees from the bottom board. Cockroaches scuttle away as I clean up the hive. There are always roaches in beehives, but a strong healthy colony keeps them in check; it is only when the colony becomes weak or dies out that they can take over. The next few colonies are strong and prosperous, and now that I have an idea of how much brood should be in each hive I disturb them as little as possible. I glance down between the frames, dust them with medicine and quickly close up the hives. The bees are barely aware that I have intruded. The neighboring hive has few bees in it when I open it, only a fraction of those that have been in the other hives. I start pulling frames out to see what is the matter. They have a queen, for in the center frame I find a patch of sealed brood, developing larvae and eggs, but the entire brood area covers a circle of a mere four or five inches in diameter, and other frames have no brood at all. A few cells contain nectar, but the bees have no stored honey. I check my record book and find this colony produced only one super of honey last year. The queen is a poor one who needs to be replaced, and after marking this down in the record book—and because I fear the colony will starve before requeening time—I fill its feeder with sugar syrup. I medicate the bees, close up their hive and ask them please to hang on.

It is silly to talk to bees—for one thing, they can't hear—but I often do anyway. I tell them encouraging things, ask them for help and always thank them for doing good work. It is said that when a beekeeper dies someone must go and tell his bees about his death or they will fly away. Whittier wrote a poem about the practice, which dates back for as long as humans have kept bees. In the West Country of England, the custom also requires tapping on the hive three times and giving the news with each tap. If this ritual is not observed, someone else in the beekeeper's family may die within a year. It all sounds very superstitious, but I like the courtesy toward bees implied by the custom; I hope someone remembers to tell my bees when I die.

When I come to the hive that had Chalk Brood, I also examine it more carefully. The Chalk Brood itself has cleared up, and the queen is laying plenty of eggs. The new larvae are developing properly. There is little sealed brood, however, and because many of the adult bees are old ones who were born last autumn, there will soon be a big drop in colony population as they die, their normal life span at an end. The decline is already indicated by the fact that stores are low in this hive. I feed the bees, medicate them and close them up.

The rest of the hives are normal. All of them have at least three frames of brood, some have more. Stores in all of them are adequate.

I carry the parts from the hive that died out back to the truck, replace my smoker in the metal carrying can, gather up the rest of my tools, fill out the record book, pull off my gloves and veil and leave. It has taken over an hour to work through these hives, but I will be able to work the other yards more quickly, because I have a good idea of what a typical hive should look like today, when the daffodils have just started to bloom.

I had planned my route so that by midday I would be at a beeyard on one of the upper branches of the most beautiful of Ozark rivers, where I like to eat my lunch. The bees are on a bluff overlooking the water, and some years ago I wedged a couple of old boards into the bluff to make a seat next to the hives. I walked to the spot carefully, because the hillside was covered with white flowers—delicate rue anemones (which have a lovely Latin name that slips easily over the tongue: *Anemonella thalictroides*), and bloodroots, *Sanguinaria canadensis,* with their orangish-red stems and flowers so tender that they bloom generously one day and drop the next. I unpacked my lunch and sat staring at the river while I ate.

This branch joins others downstream, and the cold spring-fed river it becomes is popular for fishing and floating. Last summer a friend and I spent two hot days canoeing on some of its prettiest stretches. The current is strong and paddling is not, strictly speaking, necessary, although steering is, for it is a tricky river. Late in the day we spotted a fine-looking gravel bar where we could pitch our tent for the night, but, according to our map, a regular campground was not far distant and my friend, who is a thorough man, wanted to check it out. The current swept us along and in a few minutes we pulled into the quiet creek off the river where the campground lay. It was surrounded by woods. The air was still and stifling, gnats and mosquitoes were everywhere, and camping spots were spaced at tidy, close, regular intervals. A long list of regulations and prohibitions was posted on a sign. We decided to leave our canoe tied to the bank and walk up a path we thought might lead back to the gravel bar to see if it would be a more cheerful place. The path quickly disappeared and, sweating in the heat, we slogged through a discouraging amount of poison ivy before we reached our goal. Yes, it was a beautiful place, unregulated and unruly. Wilder and cooler. We decided we wanted to stay there, but our gear was still in the canoe. We were hot, sweaty and did not fancy a walk back through the poison ivy so, with all of ten seconds consultation, we pulled off our clothes, left them in a pile and jumped into the river to float back downstream. The river was so cold it took our breath away and the current was so strong that the river seized us and tossed us about. I pretended we were otters. Almost as soon as we had plunged in, a strong rain squall began and it seemed that there was nearly as much water above us as around us. Our breathlessness, the cold wetness below, the warm wetness above and our loss of control to the river's power struck both of us as funny and we laughed as we were tumbled along by the current. In no time we were at the mouth of the stream where we had left our canoe. We swam to it and pulled ourselves out on the bank. Now we had to paddle upstream in the rain against the strong current. Our clothes were back on the gravel bar, getting wet in the rain. Our only dry change was tucked safely inside a waterproof bag; there was no point in getting another set of things wet, so it seemed only sensible to hop into the canoe in the altogether and start paddling back upstream. My friend, heavier and a more skilled canoer, took the stern. From that position it was not immediately evident that he was unclothed; after all, men often go shirtless in the summer. Women seldom do and in my position at the canoe's bow I was hard to miss. Nudity and even skinny dipping are not regarded in a kindly way in the Ozarks. The river we were on is supervised by the U.S. Park Service, and although most of my energy was consumed in paddling hard enough to make headway against the current, I did wonder briefly if nakedness was a punishable federal offense. As we got nearer our gravel bar, the rain still pelting down, the river narrowed and we began losing headway against it. We steered to the shore, clambered out and dragged the canoe the last yards upstream.

The rain squall blew on downriver, and our campsite was cool and lovely and bug free. In the evening the mists rose from the river. Before we went to sleep, we sat by our campfire and watched the moonlight reflecting on the water.

After I had finished lunch, I began work on the hives. I found one that had died out during the autumn—I could deduce that this was when it had happened, because wax moths had taken over the hive while the weather was still warm enough for them to multiply. Their larvae had ruined the comb, tunneling through it and leaving ropy webbing and frass. Tough white inch-and-a-half long cocoons were everywhere, tucked between frames and in crevices and niches throughout the bottom board, inner cover, telescoping cover and the hive itself.

There are two species of wax moth. The greater wax moth, *Galleria mellonella,* is the more common. The adult is an undistinguished, inch-long, night-flying grayish moth that may sometimes be seen flying around lights during the summer. It is worldwide in distribution, and is extremely common where bees are kept.

The lesser wax moth, *Achroia grisella,* has similar habits and distribution, but as its common name suggests, it is a smaller moth, only half the size of the greater.

Adult wax moths lay their eggs in beehives where their young will have a source of food, but in strong healthy hives the bees will kill and carry away most larvae and so keep them in check. As with cockroaches, ants or any other opportunists who would like to take advantage of the benign conditions inside a beehive, it is only when a hive of bees becomes ill or weak that wax-moth larvae can take over and destroy comb. A web-filled, wax moth-infested hive is a discouraging sight to a beekeeper, but in some ways the wax moth larvae are the beekeeper's friends, because they destroy comb from weakened colonies and keep disease such as American Foulbrood from spreading.

Today wax moths are found worldwide, for they have followed the spread of honeybees. This was not always so. Before white men and women colonized this continent, there were no honeybees in North America and there were no wax moths, either. The early settlers brought the first honeybees from Europe, and they spread by swarming ahead of the whites. The Indians, who associated them with the invaders, thought them creatures of ill omen and dubbed them "White Man's Fly." The bees were spreading into a new habitat, and like any plant or animal who moves into a fresh ecological niche, they were unchecked by predators or pests. According to early reports, the wax moths, known then as bee moths, first began making their appearance on this continent in the early 1800s. Fifty years later they were widespread and causing concern to beekeepers, who did not understand much about them even though they had been common in Europe.

Writing in a report to the Essex County (Massachusetts) Agricultural Society, one Henry K. Oliver laments the moths:

> They are paltry-looking, insignificant little gray-haired pestilent race of wax-and-honey eating and bee-destroying

> rascals that have baffled all contrivances that ingenuity has devised to conquer or destroy them. . . . He who shall be successful in devising the means of ridding the bee world of this destructive and merciless pest will richly deserve to be crowned "King Bee" . . .

Europeans had known about the link between wax moths and hive strength and health, because the two varieties of insects had been coexisting there as long as anyone could remember. Even Aristotle had written "good bees expel the moths and worms, but others, from slothfulness, neglect their combs, which then perish." But Americans had not yet learned this, and tested their ingenuity trying to prevent the moths from getting into their hives. One of the cleverest devices invented for this purpose came into use when it was discovered that adult moths only begin flying at dusk. Beekeepers reasoned that if the moths could be closed out of the hives at evening, they would never be able to lay eggs. Small gauze beehive doors, which could be closed at sunset, were made. Yankee cleverness automated the job by hitching the doors to a long, leverlike hen roost, so that the chickens could close them at day's end, when they jumped on them to go to sleep, and open them at dawn, when they jumped off to begin their chicken days. Nowhere is it recorded how well nineteenth-century chickens took to roosting on beehives, but the wax moths continued to spread.

I put the empty hive on the truck, and when I got back home I took it apart and spread it out for the chickens to clean up. Chickens may not do well as wax-moth door closers, but they do like to eat the larvae and cocoons. After a few days, when they have finished going through the debris of comb and picked out each neatly wrapped cocoon from every crevice, I'll gather up the hive parts, put fresh wax foundation in those frames that are still in good repair and then ready the hive for new bees.

I have heard that wax-moth larvae make dandy fish bait. I am not a fisherman, so I cannot report at first hand if this is true.

Honey sales business has taken me away from my farm and made me late in making the second round of the beeyards. I begrudge every day I am away from the Ozarks at this time of year, because these hills and riverbanks are so beautiful. I try to keep track each year of the succession of bloom of the wildflowers and flowering trees for the bees' benefit . . . and for mine.

Bees' vision is different from ours. Not only are their eyes structured so that they can see broken surfaces and movement easily and stationary objects less well, but their perception of color is unlike ours. I was at a state bee meeting once at which a federal bee researcher of renown, Dr. Eric Erickson, presented a slide show comparing the way we see the color of flowers with the way the bees do. Bees can see ultraviolet, which we cannot, but their eyes cannot receive the long wavelengths on the red end of the color spectrum. A flower that we call white looks blue to a bee. Brilliant ultraviolet nectar guides—straight lines pointing to the interior of a light-blue flower—shimmered deep blue in one of Dr. Erickson's slides of a bee's vision. It was striking and elaborate. But his next slide showed us the same flower as it would be seen by a human being—it was a simple, unadorned white blossom. The nectar guides visible to the bees are invisible to us. Since bees and their kin do not perceive red, plants with red flowers are bird-pollinated. (This is the reason that most humming-bird feeders are made of garish red plastic.)

I am always struck by the abundance of early spring flowers that we in our arrogant and homocentric way call white. My notebook, sketchily kept as it has been this season, lists harbinger of spring, rue anemone, pussytoes, spring beauty, toothwort, Dutchman's-breeches, bloodroot, hepatica, white violets of several species, serviceberry and wild plum, all of which have bloomed before the first week of April and all of which are white. I try to imagine I am a bee, and think of them as blue.

The relationship of the flowers that need pollinating to their pollinators interests me more than I have time to allow it: I want to think about this a great deal. I am looking forward to being old. No one could expect an old lady to keep more than a hundred hives of bees, so I'll be able to surround myself with flowers and animals. I'll have a wildflower garden, for I'll have time to create the special conditions needed for wet-loving and woods-loving plants to grow around my upland cabin. I'll watch the flowers and learn their pollinators, see how they all fit into the community of plants and animals that live here.

I have read somewhere that early bumblebees pollinate Dutchman's-breeches, and I should like to see them at work. I snatched some hours one sunny afternoon a week ago, and spent it down by the river's edge in a glen that the river had flooded and left filled with sand. The Dutchman's-breeches there were so thick that I had difficulty walking among them lest I trample some. I found a spot to sit, but all the time I was there I never saw a bumblebee. However, the Dutchman's-breeches return year after year, and so, I hope, shall I.

Mixed with the white blossoms today are masses of greenish-yellow fragrant sumac—*Rhus aromatica,* an undistinguished sprawling shrub which grows all over the eastern part of the country. To people who are not beekeepers, most members of this genus—which includes poison ivy *(Rhus toxicodendron)* and the summer-blooming sumacs with their white flowers that ripen to red clusters of berries much loved by bluebirds—are considered trash plants at best. But to a beekeeper they represent good forage for bees. Fragrant sumac, the first of the genus bees use during the season, is a source of pollen here in the Ozarks. To the casual observer, it has an alarming resemblance to its cousin poison ivy (which bees also like), with three oval-toothed leaflets. But the habit of growth is neater, more compact than poison ivy's; the leaves are smaller and dull, not shiny as poison ivy's are, and they are aromatic besides.

I allowed myself a walk out back yesterday afternoon when I returned to the farm. The air was warm and fragrant with the perfume of flowers, and I hoped I was not too late to see the blooming of what Ozarkers call fire bush. All over these hills there are foundations of long-gone cabins—overgrown, rock-lined walls that show where families lived back in the days when the big timber was cut in this part of the country. In most of them, like the one on the back of my place, there are also flowers. Iris, bridal bush, and fire bush were planted, I like to think, by women who put them there to brighten their harsh homesteading lives. The fire bushes, tough shrubs that in other parts of the country are called flowering quince *(Chemoneles sinensis)* have spread, crowding out other plants. They do not bloom for long, but when they do they fill a sunny spot in the encroaching woods with a blaze of fiery pink. I try to keep the appointment with the one on the back of my place each spring. Yesterday I did. It has grown to house size, and the masses of blossoms were filled with bees wallowing in bright-yellow pollen.

With so many flowers in bloom I expected the bees to be in good shape on this tour of my beeyards and I was not disappointed. In addition to the medicine that I carried with me today for a second and final application, I also had several jugs of sugar syrup but found no hive that needed it. What I did find, however, were several hives in which the bees were superseding, or replacing their queens. It is common for bees to do this at this time of year, and in my check of the other hives in the week ahead, I expect I'll see a number of other colonies doing the same thing. Some beekeepers believe it is a good idea to requeen their hives every year. I don't agree. Those beekeepers kill the old queen and replace her with a new one from a bee breeder. It is argued that hives with fresh queens are less likely to swarm, and that bees with a new queen will be more productive.

Aside from the decidedly cruel aspect of routinely killing perfectly good queens, I have found in the long run it is better to let the bees raise themselves new queens when they decide the time has come. They know their needs better than I do. I keep detailed records on my hives and know how much honey each one produces, and, as long as they produce well, I do not requeen them. There are many hives I have never requeened in the fifteen years since I began keeping bees. They have requeened themselves and are on their way to developing a strain of bees well suited for local conditions. I do not interfere with them, as long as they are healthy and productive, and I give them a new commercial queen only when they cease to be productive.

Queen bees, depending in part on how well they are mated, will be able to lay fertile eggs for one, two or even more years. Their only purpose within the hive is to lay eggs. They cannot even take care of themselves. The queen's attendant bees feed, stroke, groom her and carry away her feces. The workers are, as a result, aware of her condition, and when she begins to fail in health or ability to lay fertile eggs, they raise a new queen; the process is known to beekeepers as supersedure. To do this, they select a freshly laid fertile egg—one that would ordinarily develop into a worker bee—and instead of providing the larva that hatches from it with the usual worker-bee fare, they float it in royal jelly, a food rich in B vitamins which is produced from their own glands. It is this food that turns the larva into a queen. Usually, but not always, the egg selected for this special treatment is partway up a frame. The capped cell in which the queen pupates is not neatly flush with the other cells. It extends, looking rather like a peanut, down the front of the frames, where it is readily visible when the beekeeper peers down between them. It is called a queen cell.

Later on in the season, if the bees are in the mood to swarm, they will raise many queens; the queen cells are similar in appearance, but

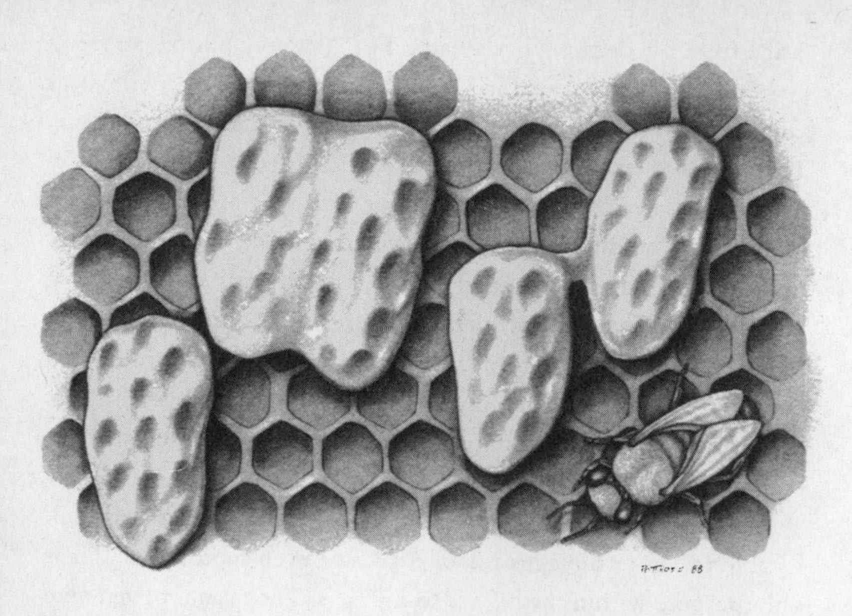

Supersedure queen cells

can usually, though not always, be found along the bottom of the frame hanging pendant below it.

When the virgin queen, a long, elegant, wasplike bee, emerges from pupation, she roams the hive to murder any other queens that the bees may have raised, in just the manner François Huber described on p. 56. She has a stinger, which she will use against other queens, but a beekeeper, if he wants, may pick her up safely in his bare hand. The worker bees will not allow her to linger in the hive because she is not yet mated, so they urge her toward the hive entrance. She then flies out on what is called her nuptial flight, to mate with the drones.

The workers and the queen develop from fertilized eggs and thus have a full paired complement of genetic material from both of their parents. They are called diploid. But the queen can also choose to lay sterile eggs—eggs that have not been fertilized and that have only half of the genetic material: genes from their moth-

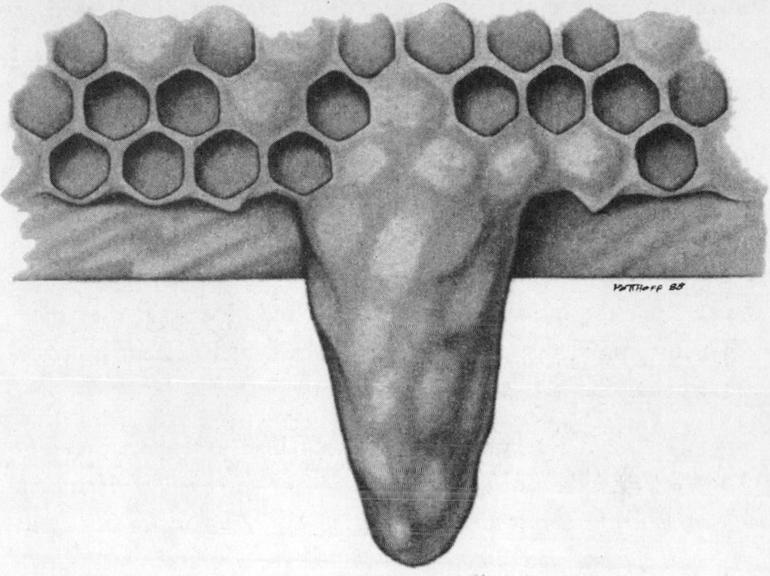

Swarm queen cell

ers only. Those develop into drones—male bees—and are haploid. Drones are bees who have grandfathers but no father, and who produce grandsons but no sons. Their only role in the colony is to mate with nubile queens. They are not physically equipped to collect nectar or pollen, nor can they defend the hive, for they have no stingers. They are big bees, with big eyes, and they hang out in groups together, watching for a virgin queen. They are not very bright; their brains are small, but not so small as a queen's, and they have been known to try to mate with a swallow flying by or an artificial queen trailed on a helium balloon by entomologists.

When a drone sees a queen, he flies high in the air to mate with her. He mates by everting his penis into her sting chamber, which closes around it, causing it to rip loose from his body as he bends over backward and falls lifeless to the ground. Queens usually copulate a number of times on their nuptial flights, and when they return to the hive they trail portions of drone gut, which beekeepers euphemistically call "mating sign."

Drones are found in bee colonies during the spring and early summer, when the workers regard them with favor. Their presence may even help to keep the hive temperature as high as bees like it during this sometimes cool season. But after the queens are mated the drones are no longer needed, and they are a drain on resources, so when the nectar flow begins to taper off in the summer the workers bar the remaining drones from the hives, and they die.

To my knowledge E. B. White never kept bees, but he wrote a fine poem about them:

Song of the Queen Bee

"The breeding of the bee," says a United States Department of Agriculture bulletin on artificial insemination, "has always been handicapped by the fact that the queen mates in the air with whatever drone she encounters."

When the air is wine and the wind is free
And the morning sits on the lovely lea
And sunlight ripples on every tree,
Then love-in-air is the thing for me—
  I'm a bee,
  I'm a ravishing, rollicking, young queen bee,
  That's me.

I wish to state that I think it's great,
Oh, it's simply rare in the upper air,
  It's the place to pair
  With a bee.
Let old geneticists plot and plan,
They're stuffy people, to a man;
Let gossips whisper behind their fan.
  (Oh, she *does*?
  Buzz, buzz, buzz!)

My nuptial flight is sheer delight;
I'm a giddy girl who likes to swirl,
To fly and soar
And fly some more,
I'm a bee.
And I wish to state that I'll *always* mate
With whatever drone I encounter.

There's a kind of a wild and glad elation
In the natural way of insemination;
Who thinks that love is a handicap
Is a fuddydud and a common sap,
For I am a queen and I am a bee,
I'm devil-may-care and I'm fancy free,
The test tube doesn't appeal to me,
Not me,
I'm a bee.
And I'm here to state that I'll *always* mate
With whatever drone I encounter.

Let mares and cows, by calculating,
Improve themselves with loveless mating,
Let groundlings breed in the modern fashion,
I'll stick to the air and the grand old passion;
I may be small and I'm just a bee
But I *won't* have Science improving *me*,
Not me,
I'm a bee.
On a day that's fair with a wind that's free,
Any old drone is the lad for me.

I have no flair for love *moderne*,
It's far too studied, far too stern,
I'm just a bee—I'm wild, I'm free,
That's me.

I can't afford to be too choosy;
In every queen there's a touch of floozy,
  And it's simply rare
  In the upper air
  And I wish to state
  That I'll *always* mate
With whatever drone I encounter.

Man is a fool for the latest movement,
He broods and broods on race improvement;
What boots it to improve a bee
If it means the end of ecstasy?
  (He ought to be there
  On a day that's fair,
  Oh, it's simply rare
    For a bee.)
Man's so wise he is growing foolish,
Some of his schemes are downright ghoulish;
He owns a bomb that'll end creation
And he wants to change the sex relation,
He thinks that love is a handicap,
He's a fuddydud, he's a simple sap;
Man is a meddler, man's a boob,
He looks for love in the depths of a tube,
His restless mind is forever ranging,
He thinks he's advancing as long as he's changing,
He cracks the atom, he racks his skull,
Man is meddlesome, man is dull,
Man is busy instead of idle,
Man is alarmingly suicidal,
  Me, I'm a bee.
I am a bee and I simply love it,
I am a bee and I'm darned glad of it,

I am a bee, I know about love:
You go upstairs, you go above,
You do not pause to dine or sup,
The sky won't wait—it's a long trip up;
You rise, you soar, you take the blue,
It's you and me, kid, me and you,
It's everything, it's the nearest drone,
It's never a thing that you find alone.
  I'm a bee,
  I'm free.

If any old farmer can keep and hive me,
Then any old drone may catch and wive me;
I'm sorry for creatures who cannot pair
On a gorgeous day in the upper air,
I'm sorry for cows who have to boast
Of affairs they've had by parcel post,
I'm sorry for man with his plots and guile,
His test-tube manner, his test-tube smile;
I'll multiply and I'll increase
As I always have—by mere caprice;
For I am a queen and I'm a bee,
I'm devil-may-care and I'm fancy-free,
Love-in-air is the thing for me,
  Oh, it's simply *rare*
  In the beautiful air,
    And I wish to state
    That I'll *always* mate
With whatever drone I encounter.

This part of the Ozarks is on the edge of a major flyway for migrating birds, and although few of the summer residents are here—I've yet to hear indigo buntings singing, for instance—the warblers have begun to make their appearance. For most of them, this is a way stop only and each spring I have to relearn the various kinds. But for the past several days I have been hearing a call I know well. It is the small Bronx cheer of the blue-winged warblers, who live here all summer. Blue-winged warblers are canary-sized, canary-yellow birds with blue-gray barred wings, olive backs and dressy-looking black lines that run from their beaks to the back of their heads. The guide books hold that blue-winged warblers say *Bee-Bzzzz.* It would be more accurate to say that when you pass one hidden in the low shrubs, you have the distinct impression that you have just been given the raspberry. I haven't had much time to spend watching warblers, however. They arrive at the same time I begin my work with queen bees.

Every winter, I place an order for mated queens with a southern bee breeder and try to guess what the weather will be like in April, when they will be shipped. The queen work is exacting, and I need sunny mild days to do it easily. I usually order fifty queens each year, and ask that they be shipped in two batches one week apart; even with the best of weather, it will take me an entire week to establish twenty-five queens in new hives, and a few days of rain or cold will not only throw off my schedule, but may result in the death of some of the queens. The queens are raised in southern apiaries and captured there in their hives on sunny days. Each one, with a few worker bees to attend to her needs, is put into a three-inch long wooden cage faced with a wire screen. One end of the cage is plugged with sugar, which the bees will use for food, the other with a small cork. The cages are stacked carefully in a neat bundle with the screen opening against the next wooden cage back. If the screened openings faced, the queens would sting one another to death. Separated, they are still aware of their rivals and fuss at them. The first thing I do when I receive my bundle of twenty-five queens is to pry apart the package and check to see that all the queens are alive and active. As I spread them out on the kitchen table I can hear them testily challenging one another in a shrill, high-pitched *ze-eeeep zeeee-ep.*

The queens are shipped airmail, but even a few days in the mails are too many for a bee, so I try to get each queen into better quarters as quickly as possible, lest her health or egg-laying abilities suffer. For the past several days, I have been assembling the equipment I'll need to make up fifty new hives. Each one is made of a bottom board, a single hive body, an inner cover and a telescoping cover. I staple the bottom boards to these hive bodies, because each nucleus hive—or nuc, as beekeepers call it—will be moved from place to place. Inside I put a feeder and eight good frames of fully worked comb, and I screen the entrances.

Every year, I make up some hives of bees to sell to people who

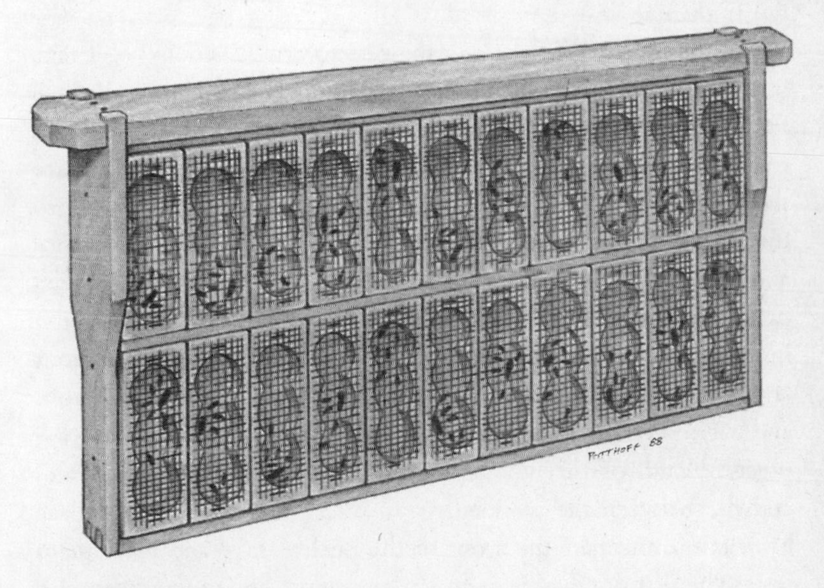

Queens in mailing cages ready to go into temporary queenery

want to get started with them in the springtime. I also need a few new hives to replace ones that have died out during the previous year, and I need some additional ones for requeening. I use the same procedure for all three needs. There are a number of other ways to requeen, to be sure, which do not involve making up a starter hive but instead call for placing the queen directly into a colony that needs a new one, and then practicing a variety of subterfuges to get the bees to accept her rather than surround her in an angry mass and suffocate her, a process known to beekeepers as "balling." I have tried those other methods with varying degrees of success and numbers of dead queens. But I have never had a problem with getting bees to accept a new queen when I took the extra step of first starting a queen in her own new temporary hive. It is a procedure that works with the bees' own biology and behavior rather than against it.

On the morning of the day the queens are due to arrive, I take one of the nucs out to the hives in my home beeyard and pull three frames of brood and young bees from one of them, having first found its queen and made sure that she stays with her hive. I replace the frames I take with empty ones and put the brood frames into the nuc. Then I pick up my shipment of queens from the post office and check them at the kitchen table. I press each acceptable queen cage into a queen frame, after I have assured myself that the queen inside is lively and well. A queen frame is an ordinary frame without any beeswax in it but with two thin strips of wood running from end bar to end bar. These wooden strips divide the frame into the precise length of the wooden queen cages. I put the queen cages endwise between the wooden strips, where they are held in place by friction, and take them out to the nucleus hive and place them in it. The first nuc has become my queenery. The young bees who emerge from the brood area will have no loyalty to their former queen, and small groups of them will immediately start to tend the queens in each of the wooden cages, poking their long tongues through the screened openings to feed them, and covering the screens to generate enough heat to preserve the queens during the cool spring nights. The queenery still does not represent natural conditions for queen bees, but it is better than what they have had for the past several days in the U.S. mails and will serve to keep them in as good condition as possible while I prepare permanent homes for them.

The queenery system works well enough in most weather, but one spring a late snowstorm blew in just as I had finished making up the queenery. I watched as the temperature dropped alarmingly throughout the day and listened to the forecast predicting a 10° F. night. I knew the small force of bees would find it impossible to keep the queens warm in their cages, so I brought the queenery inside my cabin and wrapped it in an old bedspread to cut out the light from the bees, who would try to fly toward it. They were quiet during the night, when it was dark anyway, but when light began to come through the eastern windows, some bees, made lively from the warmth of the woodstove, began to find their way out through the folds in the bedspread. They flew about the house, confused by the interior landscape—which was thoroughly unsuitable from a bee's standpoint—and clustered in small dispirited groups on the windows. The weather stayed cold and blustery for several days and the bees and I remained roommates for all of them. A friend stopped over one afternoon for coffee, but excused herself when she saw the bees flying around. "I think you may be carrying this bee thing a little far," she said, backing out of the door. I missed having coffee with my friend, but I saved the queens. Outdoors, each and every six-dollar queen would have died; there were not enough bees to warm the chilled air spaces inside the mailing cages.

This year, fortunately, the weather has been more favorable for bee work, and yesterday, after I had put the twenty-five queens in their temporary queenery, I was able to go directly to the beeyards and start preparing better homes for some of them.

Queen work is the most exacting part of beekeeping. When I make up nucs I must start with three frames of brood from an established hive, and in order to obtain those three frames I must first locate the queen in the hive I am taking them from to make sure she stays behind. If I were to accidentally put her in the nuc, not only would the parent hive be queenless, but the old queen would kill the new one. The hives now, during warbler time, can contain more than forty thousand bees, and finding the single queen bee can be tricky. It seems daunting to a new beekeeper, but fortunately, with practice, it becomes easier.

Yesterday I looked for queen bees in one of my closest beeyards. It is at the edge of a woods near an old houseplace on a hill. I drove through a cow pasture to get to it, backed up the pickup as near to the hives as I could and unloaded all my empty nucs. I opened the first one and took out all the frames, leaving the feeder in place. Then I removed the lid of the first hive in the row, putting the lid upside down on the ground. I had left the inner cover in place, and carefully pried loose the top hive body from the bottom one with my hive tool. Then I set the top hive body catty-corner across the telescoping cover. If the queen were in it and were to fall off, she would still be on the inverted cover and not on the ground, where she might be lost or get squashed. Now that I had the bottom hive body exposed, I took out the plastic trough feeder and turned it over, checking to see that the queen was not on it, walking around among the worker bees. She was not; I really hadn't expected her to be. I set the feeder to one side and began taking out the frames, starting with the outermost ones on either side. The frames themselves were empty; the queen was not on them. I set each one beside the feeder on the ground next to the hive. The bees were in such a good mood that even taking apart their home failed to make them angry. The day was sunny, the air was warm, flowers were in bloom everywhere. Most of the older bees—the ones more likely to express hostility about a disruption to their hive, were out foraging for

Upper hive body on cover, showing empty nuc and bottom hive body, with combs leaning against them

nectar and pollen. When they returned, many of them with yellow, crimson or orange wads of pollen on their hindermost legs, they acted confused. Their hive was in a shambles, their frames spread outside it. They still were not cross, but flew through the entrance or the remaining ventilation hole and crawled up inside, appearing bewildered by the unaccustomed condition of their house. Some of the bees from the top hive body had flown back and were circling above the bottom hive body. They seemed disoriented, but they were not aggressive. I had puffed a small amount of smoke into the hive when I first opened it but I had not needed to use any more.

Even though their mood was good, I tried to work as quickly as I could to bother them as little as possible. I continued to check through the frames, looking for the queen. I pried each frame loose with the hive tool, and then lifted it out with the grips, carefully checking one side and then the other.

The queen is longer and more slender than either worker bees or drones. She walks in what can most accurately, if anthropomorphically, be described as a purposeful way, so the best technique to use looking for her is that of the gestalt. I try to take in the entire frame side at a glance, relying on the eye to pick out the anomalous shape and motion of the queen among the hundreds of bees rather than looking at each part of the frame separately. As one learns to shift from sequential focus to a state of open and alert awareness, the eye gradually gets better at spotting a queen.

The queen was not in the bottom hive body, but I did find a few frames of fresh nectar and pollen there, and, in between them, two frames of eggs and larvae. I set the nectar frames aside and returned the frames with developing brood back to the bottom hive body. As long as I was taking apart the entire hive, I wanted to fill the bottom hive body with as many brood frames as possible, leaving the empty ones above—a tactic that might deter this hive from swarming later on.

The queen had to be in the upper hive body, the one that straddled the inverted cover. I examined the inner cover to see that she was not walking around its underside, then began removing frames, starting with those on the outside. There were many more bees in the upper hive body than in the lower, so I had to be more careful that I didn't miss the queen. On cold or windy days, the bees bunch and cluster on the frames as soon as they are pulled out, and on such a day the queen could be in the center of any one of the clusters. As a result, when the weather is poor she is usually impossible to find. But yesterday there was no wind, and it was warm enough for the bees to be spread no more than one bee deep on the frames. The top hive body still had last year's honey in its outermost two frames, and once I was sure the queen was not on them, I put them on the ground beside the ones from the bottom hive body. The seven remaining frames all contained brood, and it was on those I expected to find the queen. She was not on either the first or the second, but I did find her on the third. She was walking along sedately, looking for an empty cell in which she could lay an egg, and was accompanied by the devoted group of workers who follow her everywhere, tending to her needs.

I carefully put her frame into the bottom hive body, nestling it in with two others so that she wouldn't be squashed when I reassembled the hive. This hive had plenty of brood to spare; in fact, its general well-being will be improved by taking these three brood frames from it, because it will be less likely to swarm. I placed the three, with many young bees clinging to the sealed brood, in the open nucleus hive, and added one of the frames of honey and pollen to them before filling it with combs from the bottom hive body, all of which had young bees on them. The nuc already had a screen stapled in place barring its entrance, and after I put its inner cover and telescoping cover in place, I taped the ventilation hole shut to secure the bees inside. Then I carried it back to the pickup.

I reassembled the donor hive, giving it mostly empty comb in its second story from the supply that had been inside the nuc. The bees were still confused about the terrible disruption to their quarters, but their mood stayed good. As I closed them up, I could see them bustling about setting things to rights.

I was able to make up seven more nucs from the hives in this yard, even though in two of them the queens had hidden themselves so effectively I was unable to find them. I searched through the combs twice and then gave up, because after a second pass the bees are so completely disturbed that the queen could be anywhere; it simply is not worthwhile to try to find her a third time. But the eight nucs took up all the available space in the bed of my half-ton pickup anyway.

When I arrived back at my own farm, I unloaded the nucs along the roadway to my woodlot, where they would be easy to move again. Then I untaped the ventilation holes, and the bees, out of sorts from the upheaval, flew out and hung in the air in a confused way, trying to make sense of the unfamiliar new landmarks. Some of them were beginning to be cross, and would have stung me had they been able to get through my protective clothing.

I went out in back to my home beeyard, and opened the queenery. I took from it eight queens in their cages and returned to the line of nucs. I smoked the eight to quiet them, and then began handing out their new queens.

I opened the first nuc, smoked the bees down within it and selected one of the queens in her cage. I removed the cover on the sugar-plug end of her cage and poked a hole through the sugar with a nail to make the bees' job of chewing through it easier. Then I wedged the cage, screenwire-side down, in between two of the inner frames and closed the hive. I repeated the process in the seven other nucs. My part of the work was over. From then on it would be up to the bees.

These bees are cross, and have found nothing agreeable in the day's events. Many of them still carry a chemical memory of the queen from whose hive they were taken. If I were to open up the queen cage by taking out its cork and give them their new queen directly, they would ball her. As it is, they can't get at her because the screenwire intervenes. She is of great interest to them, however. Over the next hours and days they will become acquainted with her and her unique pheromone through the screen as they feed her. In addition, they will chew through the sugar plug at her cage's end, which, in warm sunny weather, may take only a day. Once the plug is gone, the bees are free to get inside the cage and the queen is free to come out. By that time, their hostility should be gone and they should cherish and cosset her. If she has been well mated, and if she has not been harmed in shipping, she will start laying eggs immediately after she comes out of the cage. Sometimes, although to human eyes she looks perfectly fine, the bees will discover some flaw in her, and after she has been free in her own hive for a short time will kill her and raise themselves a new queen from one of the eggs she has laid. Beekeepers call this "premature supersedure," and it is impossible to predict. I seldom see it, because I order my queens from one of the most reputable of queen breeders, but I know from talking to other beekeepers that premature supersedure is often a problem, particularly with hybrid queens.

Within the species *A. mellifera* there are a number of races of bees. Those most commonly used in this country are the Caucasian and the Italian. Caucasians tend to be a bit conservative, holding back their spring buildup until the weather is stable, chinking up their hives thoroughly in the autumn to keep out Siberian winds. The Italians—the ones that Lorenzo Langstroth imported and developed—have more exuberant, sunny Mediterranean dispositions. They do not propolize so much, which makes their hives easier to work. They build up rapidly in the springtime (and sometimes starve if the weather turns poor later on and there are no flowers to sustain the burgeoning population). A single sunbeam and a lone wildflower mean springtime to an Italian bee.

Bee breeders have developed many expensive hybrids. They have created strains that are desirable in various ways—color, productivity, gentleness—and for various climates. I experimented with a number of them and found that the hybrids did not supersede truly and that others were not suited for the erratic climate in the Ozarks. I have decided plain Italians are the best for my purposes, and the strains that develop from the supersedure daughters of Italian queens seem admirably suited for local weather conditions. But every beekeeper must experiment for himself to find the breed of bees proper for his methods, his needs and particular climate.

It takes the bees in the nucs several days to settle down and become familiar with their new location. Bees learn precise maps of their territory, taking their bearings and reinforcing their sense of them when they fly out of their hives. I try to disturb them as little as possible during the several days it takes them to learn this new place and while they are at the important business of deciding whether or not they will accept the queen I have given them.

In the meantime, I have been collecting brood and making up nucs from other beeyards. I have emptied out the queenery, giving the twenty-fifth queen to those queenery bees to have for their very own.

After three sunny days the first group of eight nucs seems to have settled down, and I begin checking them to see if the queens have been released from their mailing cages and if they have started their egg-laying careers. In the first one I open, I find the queen still in her cage, and discover that the sugar plug is caked and hard. I enlarge the opening a bit, and return the cage. When I open the second, I am certain all is well. The queen cage has forced the frames apart beyond a bee space, and a small piece of snowy white comb is pendant from it. It is only when bees' morale is good (and nothing makes their morale better than to be queenright) that they build comb. Sure enough, the queen cage is empty. I set it aside, and pull out one of the center frames. I don't see the queen, but I don't need to. Every empty cell contains a glistening white egg—the queen has been accepted, and she is hard at work. I replace the frame, adjust the spaces between the rest of them so that the bee space is respected and close up the hive. I put a stone on top of it to remind myself that it is ready to be moved out, used for requeening or prepared for sale, and continue to check these first eight, all of which are ready except for the one in which the queen had remained in her cage.

This past winter, I have taken orders for ten new hives of bees from people who wanted to get started with an established hive in the spring with as little risk and work as possible. In a few days, I have ten nucs in which the queens have a good egg-laying pattern so I bring ten more hive bodies out to the row. I feed each of the ten a full ration of sugar syrup and put a second hive body in place on each one. Every other day over the next several weeks, I'll feed those ten hives again. At the end of that time, the queen, responding to the bees' extra feed, will have moved up and begun laying eggs in the upper hive body. The bees, aided by the combination of extra feed and plentiful flower bloom, begin to store nectar and pollen there too.

When I decide the new young hives are strong enough to manage on their own, I telephone the customers who have bought them and we arrange an evening convenient for both of us when they can drive out to pick up their bees. The hives are all stapled, and I have left the moving screens in place on each one. When the new owners arrive at dusk after the field bees have returned home, all that remains to do is to tape shut the ventilation holes and help load the hives on the customers' pickups.

These customers are people new to beekeeping, and need advice. I tell them to feed the bees a time or two more while the bees are learning yet another new location, and to watch the weather and flower bloom in case they need to be fed later on. I recommend a book about bees to read. I tell them how to set the hives up on boards with a forward tilt to allow moisture to run out. I advise them to face their hive toward the southeast. I tell them the bees will need shade and water to drink and for cooling. New beekeepers usually have questions, and we often talk bees on into the evening.

Last spring, I began listening to myself talk and noticed what I sound like. I sound like a mother relinquishing her firstborn to the kindergarten teacher. I sound like a writer handing her manuscript to her editor. I sound like a Republican tax assessor turning over the job to a Democrat.

I don't need to add hives, I remind myself. My bees already produce more honey than I can sell. I can't keep every colony of bees I start. I can't, but I want to.

The second batch of twenty-five queens came into the local post office on Tuesday, which meant that they had spent the weekend in the mails somewhere, and after I picked them up I worked fast with them. When they are temporarily imprisoned in cages they are as dependent on me as any domestic animal might be for its well-being. Although we interfere with bees' breeding, talk of them as though they are domesticated and keep them in manmade hives, bees are wild animals and, like any wild animals, need to be free to live. Strictly speaking one never "keeps" bees—one comes to terms with their wild nature. I had to restore their freedom and independence as soon as I could.

The week the second batch of queens comes is always the busiest of the spring for me. I have to take brood from established hives to set up nucs, as I did the previous week, and the second shipment takes more time, because the hives are more populous and the old queens are therefore harder to find. In addition, I have to move the nucs established the week before into permanent locations.

My goal each spring is to have all hives in place at least half a week before the blackberries bloom and give the first major nectar flow—usually in the early part of May. In other parts of the country, the first strong nectar flow will come from other flowers and a beekeeper must learn to recognize whatever plants produce it and have his hives well established by the time they bloom.

The week has been a frantic one. I have been starting work earlier in the mornings than usual and staying out in the beeyards later than it is entirely comfortable to do with bees, working long past the time when the field worker bees have begun to come home for the day. I have not had time to go to the grocery store, and my refrigerator is empty. There are piles of dirty socks, and bee suits lie where I have stepped out of them. Late in the evening, friends telephone and say, "Where *have* you been? I've been trying to reach you all day," and then kindly invite me to dinner because they suspect I am not eating properly.

In the course of opening hives out in the beeyards to take brood for the new nucs, I found a queenless colony in which the worker bees had started to lay eggs. "Laying workers," as they are called, are a curious biological phenomenon. They seem to be one of nature's mistakes.

In what is called a "queenright" colony, the queen is constantly being fed and touched by worker bees, and in the process she passes on to them her own unique chemical marker or pheromone which is called "queen substance" by beekeepers. In turn, the worker bees feed one another and pass along traces of this pheromone as they share out food. The feeding is continuous and instinctive. A bee with an empty crop approaches another and thrusts her proboscis between the mouth parts of her sister. This begging behavior triggers a donor response in any bee who has food in her own crop, and she will immediately start to feed her sister. Both bees stroke one another's antennae in the process. One researcher, J. B. Free, found that a bee's antennae were the most important part of her anatomy and essential in making bees feed one another. Bees who have lost their antennae are less often fed. Free also found that donor bees would try to feed freshly severed bee heads that had intact antennae, or even small balls of cotton with tiny wires sticking out of them that looked vaguely like antennae, to other bees.

This constant feeding is a form of chemical communication, and one of the messages it allows the queen to say is "Queen here. Queen here. Queen here." As long as she keeps "talking," the ovaries of the worker bees stay undeveloped. But if something happens to the queen bee, the inhibition on the workers' egg-producing organs is gone.

Some disaster had taken the queen in the laying worker colony I found. Perhaps the bees had raised a supersedure queen to replace a failing one, and a swallow or kingbird had snatched up the virgin on her mating flight. Whatever it was, the hive had no queen and no eggs fresh enough to raise another. When this happens, a colony usually dwindles and dies out, but sometimes, for reasons that are poorly understood, the ovaries in one or several worker bees begin to develop.

These worker bees have never been on a mating flight and have no instinct urging them to do so, but the eggs within their ovaries ripen and they lay them in the cells of comb within the hive. Because they are not queens, they are not very good at laying eggs. They often lay several in one cell, then skip a few cells and lay again. They are shorter than queens, and so they often lay their eggs against the side of the cell rather than at the bottom. And of course, since they have never mated, the eggs are sterile, contain only half the required genetic material and will develop into drones.

When I discovered this hive a few days ago I was sure it was a laying worker colony as soon as I opened it, because there were entirely too many drones in it. During this time of year there are always some drones in every hive, for the queen can choose, when she wants to, to lay unfertilized eggs that will develop into drones. In the spring and early summer, they are regarded in a kindly way by worker bees, and are free to drift from hive to hive. They often make my job of finding the queen more difficult, because my eye has learned to pick out the anomalous shape on each frame and it immediately registers the big, fat, furry drones who are a distinct and small minority in most hives. But in this laying worker hive they were not an anomaly. They were in the majority.

I took out several frames and found eggs scattered haphazardly throughout them. The bigger drone larvae will hatch from these eggs. When the larvae pupate, the sealed cells bulge convexly from the surface of the comb in sharp contrast to the sealed worker brood, whose cells, in a queenright colony remain flush with the face of the comb. But there was no worker brood present in this colony at all. All sealed cells bulged outward and promised more male bees to come. There was no honey left, and little fresh nectar. Worker bees normally live about six weeks; without new young ones to grow up and replace them, the dwindling numbers of aging workers in this hive had to work harder and harder to support the increasing number of drones. It is difficult not to look at all this in anthropomorphic terms—to feel sympathy for the dutiful workers

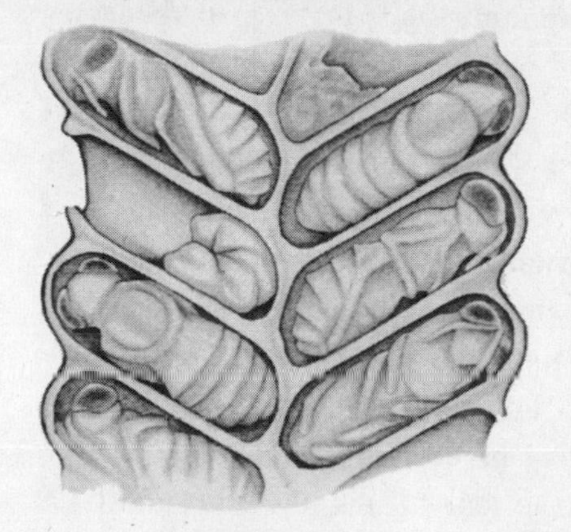

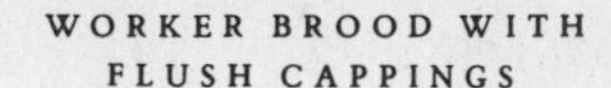

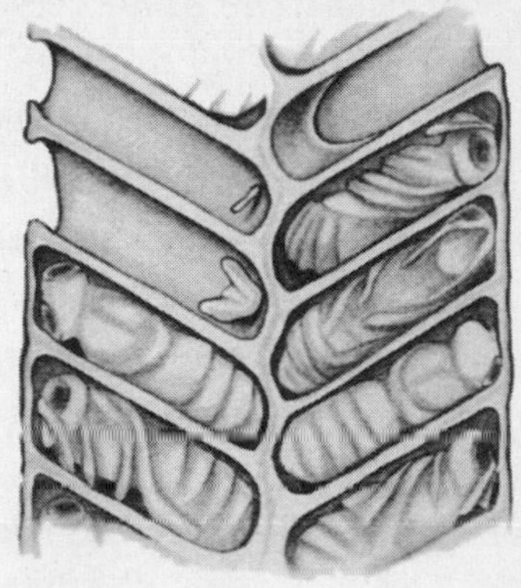

Worker and drone sealed brood

and a twinge of resentment at the greedy drones. This would be a mistake. The morale of the worker bees in a laying worker colony is surprisingly good. The remaining bees, despite the fact that their colony is doomed, are as contented with their laying workers as they would be with a queen. They will ball and kill any real queen a beekeeper tries to give them. Beekeepers consider them impossible to requeen and the usual advice is to dump out the last bees and take away the empty hive, but I have found a way to take advantage of this high state of morale and to create a productive hive out of it.

When I find a laying worker hive, I return as soon as possible, the next day if I can, with one of the queenright nucs I have created. When I get to the beeyard, I first move the laying worker hive a short distance away. Then I put the queenright nuc in its place, untape its ventilation hole and unscreen its entrance. The bees in the nuc are confused. They fly around, wonder where they are. The field bees from the queenless hive are also confused, but they return to their old location with loads of nectar for their drones, and any bee, no matter what her chemical identity, who flies into a hive heavy with nectar during fine weather is rarely challenged by guard bees, especially those who are unsure of their new surroundings. I leave the nuc undisturbed for a few minutes while I take apart the laying worker hive. I try to disrupt them thoroughly, to make them less sure of themselves on this their home turf. Some frames still have bees clinging to them, and I rap them smartly on the ground to dislodge the bees. Startled, they fly up in the air and gradually, a few at a time, go back to the original location where the queenright nuc now sits. Their morale is momentarily so reduced that they will not harm the queen. I put together all the remaining frames that contain nectar and drone brood. I consolidate them into a single hive body, and give it to the nucleus hive as its second story.

I alter my records to show the state of this new hive, and leave the bees to sort themselves out. I have found that these hives always tend to be extremely productive. The nucleus hive is reinforced by the remaining worker bees from the laying worker colony, who maintain a high morale. The drones from that colony will be tolerated by the bees in this outyard for a month or so. Some may mate with virgin queens and others, after their time of usefulness is past, will be excluded from any hive they try to enter. I have given these bees a beekeeperly nudge and put them back on their proper biological track. A hive that would have died out has been encouraged to assimilate with another, and both will prosper more than had each remained apart.

All fifty queens are now in nucleus hives. Some of these hives, such as the ones I'm preparing for customers, have even graduated into their second stories, or have been used to set up permanent hives in the outyards. But there will be no letup in my work until I get all of the bees established permanently. And my job is further complicated because there has been a pause in the bloom of wildflowers of interest to bees. The understory of my woodlot is filled with dogwood in thick white bloom, but, although they please me, dogwoods are ignored by the bees, having no nectar to give them.

The roadsides are covered with violets in all shades of blue and purple. My beeyard down on the river branch is carpeted with one of the violets' cousins, johnny-jump-ups, but bees work none of the flowers in the violet family, which is self-fertilizing. I have been keeping my eye on some of the hives that built up quickly in the early part of the spring, because now they are crowded with bees and have an enormous need for food. In a week or so, the blackberries will be in bloom, and there will be plenty of nectar for all, and enough left over to make the honey I'll be able to take at summer's end, but unless I feed some of those hives, the bees will die before the blackberries are in flower. When I stopped at the feed store yesterday, the man who works on the loading dock told me his single hive of bees had just died from starvation. He had not realized there was so little in bloom that bees could use.

So I have been carrying jugs of sugar syrup with me as I move the new hives into the beeyards. I feed the nucs when I set them out permanently, to give them a start while the bees are taking their bearings, but I also have been checking the hives already in the beeyards and feeding those that need it.

Last evening I taped shut the ventilation holes of a truck load of nucs to move today. This morning, just as the sky was turning light, I loaded them into my pickup—these small, one-story nucleus hives are easy enough for one person to lift. I stacked additional empty hive bodies on top of them to make second stories for the hives that would need them, and back at the cabin mixed up a batch of sugar syrup. I loaded the jugs on the truck, too.

Just after the sun came up, a white-eyed vireo began to sing—the first I've heard this season. Their song, which my bird book lamely describes as *chick-a-per-weeoo-chick,* once learned is impossible to mistake for any other, and I love hearing it, because it means that springtime is really here. We still have "blackberry winter" ahead of us. This is a few days or even a week of unsettled, sometimes cold weather that usually comes about the time the blackberries bloom, a last reminder that steadiness is not a feature of the Ozark climate; but the days are growing longer, the sun heats even these hills for a greater period each day, and the white-eyed vireos are guarantees of warm weather on its way.

I took my truckload of nucs to two beeyards. In the first of them, I needed one as a replacement for a hive that had died out during the winter and one more for requeening. After I drove in, I put the replacement hive down: I set out the boards for it, and put the nuc on them. I then untaped the ventilation hole, unscreened it and opened it up. After filling the feeder with sugar syrup and giving it one of the hive bodies I had brought along as a second story, I closed it up and wished the bees happiness in their new home. From a human standpoint, this is a pretty beeyard. It is on a knoll shaded by big oak trees, with a thicket of wild honeysuckle behind them. The bees don't make use of the honeysuckle, which has corolla tubes too deep for a honeybee's tongue, but they will find useful the Ladino clover the farmer has planted in the pasture in front of their hives. This is one of my best beeyards because of that clover; the new bees should prosper here.

By this time of year I have a good idea which hives should be requeened. I have kept a list of those that need new queens—either because they were unproductive last season or developed queen problems during the early spring. The hive in this yard that needs requeening has a poor queen. Last year, her hive produced only part of one super of honey. The colony did manage to make it through the winter, but she had not laid many eggs so far this year, and although all the other hives in this yard are full of bees, the population in her hive was a small one.

I set the hive to one side and put a new nuc in its place. I untape the ventilation hole in the nuc and take away the screen from its entrance, so that the field workers from the old hive can begin to mingle with this strong, feisty new group of bees. Then I take apart the old hive to find the queen. I have to find and kill her, because if I were to unite the two hives without doing so she might kill the new queen. There is little brood in this hive and the morale of the bees tending them is not strong. They are mild-mannered to the point of spiritlessness and make no objection to my taking apart their hive. In such a small colony the queen is not hard to find and when I do, I kill her. Then I put together in one of the hive bodies all the frames that have brood or nectar and pollen in them and give it to the nuc as a second story, after I have filled the feeder with sugar syrup. The morale of the original hive is so poor that they will not kill the new queen. This is as near to a foolproof system of requeening as I know. I have been using it for a number of years and have, so far, never lost a colony of bees when I did so.

I became convinced some years back, when I helped a beekeeper who insisted on using another method, that it was a chancy business to introduce queens directly into hives without first building up a nucleus population loyal to them. This is tornado country, and during springtime that year a twister ripped through the farm of a man who had just bought a hundred hives of bees as a sideline to his hog operation. When I asked him if I could help him and his family in any way, he said he would be grateful if I would help him requeen his hives the next day. He had been convinced by an article he had read that his hives should be requeened each year with hybrid queens. The day before the tornado struck, all one hundred queens had arrived in the mail, and while he and his family attended to the most immediate tornado damage, the queens had been left sitting in their mailing cages on top of his refrigerator where he hoped they would keep warm. They had been up there for a week before he could get to them.

He insisted we find the old queens, kill them and then directly give the hives the new queen; still in their cages and with the sugar plugs pierced. I told him I thought it wouldn't work well. I had tried doing it in the past, and found that well-established, fully populated hives often killed queens that were introduced so quickly.

In addition, these queens had been in their mailing cages too long and might have been harmed by their confinement. I argued as persuasively as I knew how that we take the time to establish nucleus hives first to see which queens would survive, and only then use the nucs for requeening. But he was in a hurry. He could spare just one day for the bees, he said. And he was sure his method was the best one. The poor man had had so many other worries and was so insistent we proceed that, against my better judgment, I agreed, and we began opening hives. By noon both of us were exhausted. We still had more than half the queens unhoused, so we called another beekeeper to come help us. We finished late in the afternoon. The farmer was effusive in his thanks, but although I was glad I had been able to lighten his cares, I did not think I had done a good day's work by the bees.

I met him some months later and asked how the requeening had worked out. He shook his head sadly. Not even a quarter of the hives had accepted the queens. A few of the remaining ones had raised new queens of their own, but most had simply died out. He had lost not only most of the expensive hybrid queens he had bought but a great many of his hives of bees as well. A year later, discouraged with beekeeping, he sold the remaining hives.

After requeening the hive that needed it in my beeyard, I check the other hives to see if any need feeding. Several do; they are the most populated of the hives, and the bees have no stores at all left in their combs. One of them has even begun sacrificing the male bees, for I can see dead drones strewn on the alighting board. I feed the hives and fill out my record book, noting the additions and changes I made in this yard and the hives I fed. I will have to keep a close watch on what is in bloom, and if the blackberries do not come along in a week some of these hives may need to be fed again.

In the second yard I repeated the process, requeening four and putting out four nucs as additional hives.

This second beeyard is a wilder place than the farm to which I went first. It is on a piece of land a farmer owns but does not use—a weedy abandoned spot, and the wild things are reclaiming it. The bees and I like it here, and I was pleased to see fat white flower buds on the dewberries—prostrate blackberries, which are among the first of the genus *Rubus* to bloom. The ground is too damp to sit on and have lunch, so I sit cross-legged on top of one of the hives to eat and watch bees fly out searching for flowers. Although there are not as many of them as busy and intent in their flights as there will be in a week or so, I can see that some of them are returning heavy with nectar; they are finding flowers in bloom somewhere, and so after I have drunk my coffee, I climb off the beehive and walk in the direction from which the bees returned. Back in the scrubby undergrowth I find a patch of wild black-raspberry canes. Those black raspberries are another early-blooming member of the *Rubus* genus. In two months, their fruit will be delicious to pick and eat, and will be all the more enjoyed because it will be a rare treat. There will be only a handful here and there. Not enough of them grow in this part of the Ozarks to provide a good supply of wild fruit for human beings or a major nectar source for bees, but today every greenish-white flower on them has a bee in it taking nectar. It is a welcome sight. I stand and watch and, once again, I hear a white-eyed vireo sing.

All the resident summer warblers have come back. They are setting up shop and tending to nest building. Today I lingered over my morning coffee and listened to them, especially the yellow-breasted chats, the antics among warblers. They are big as warblers go, with solid yellow breasts and olive-green backs. Their vocabulary is a series of squawks, rasps, trills and verbal punctuation marks, which they sputter out as they caper from tree branch to bush to branch again. There were two of them over near the chicken coop; I was watching them through my binoculars and listening so intently to their calls that I nearly failed to notice that I, in turn, was being watched. A great crested flycatcher, normally a shy bird, was perched on the back of the wicker chair next to the bench on which I was sitting. They are dull-colored birds with a wash of yellow on their bellies, and they usually perch so high in the trees that the modest crest which gives them their name is barely noticeable. But this bold individual was watching me calmly and seemed to find me as interesting as I found him.

We stared at each other for the longest time. Eventually he had enough of me and flew away and I, too, left to super the bees.

I have been supering now for over a week, to catch as much good blackberry honey as I can from the bees. I have been carrying truckloads of empty supers around to my beeyards all week and today was the last of that work. By day's end, every one of my hives had a super on it, and the bees were hard at work in them.

The supers each contain nine frames of drawn comb that the bees had filled with honey last year. After I had extracted the honey from them, I had left them outside to let the bees clean up the residue in the combs. Then I had fumigated the stacks of supers to keep wax moths from destroying the combs within and stored them for the winter. They are now ready to be filled again.

Some queens will go up into the supers after they are put on the hives and begin laying eggs in the combs there, but beekeepers prefer the bees to fill the supers with honey in a place that is convenient to harvest from the hive. They want the queens to lay their eggs down below, in the hive bodies. To prevent queens from getting into the supers, beekeeping supply companies sell "queen excluders."

Queen excluders are rigid welded wires placed in wooden frames cut to the outside dimensions of the hive body and supers. The wires are close enough together to keep queen bees from going through, but far enough apart to allow passage to the smaller worker bees. When in place on top of the second hive body and under the supers,

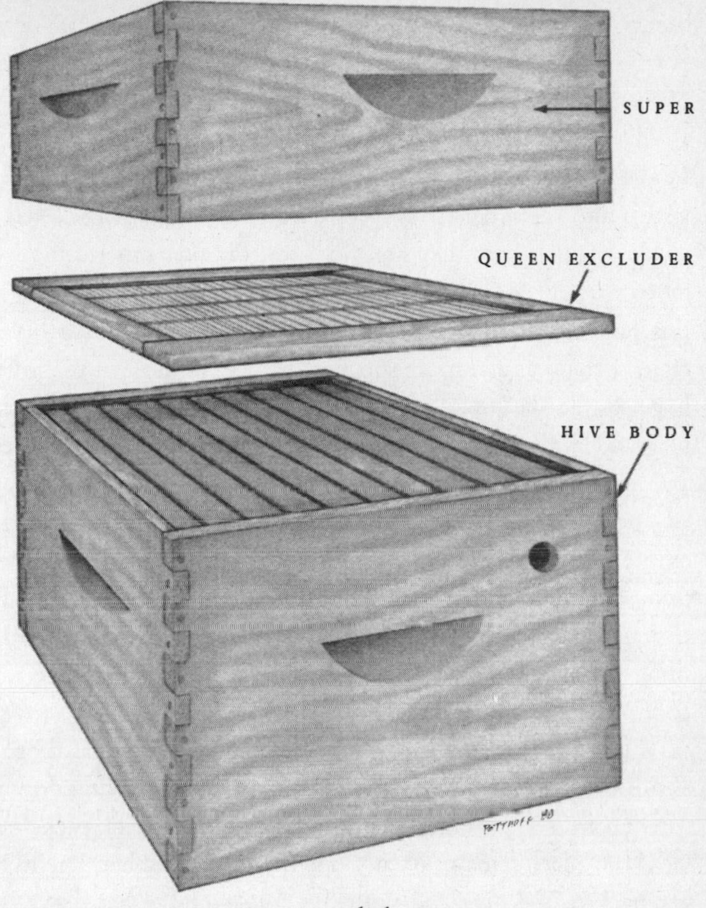

Queen excluder

they confine the queen to the lower parts of the hive and let the bees still store honey up above in the supers.

Queen excluders are controversial pieces of equipment among beekeepers. Although they do keep the queens tidily below, the worker bees are also reluctant to go through them. I know a commercial beekeeper with three thousand hives who says they are not only queen excluders but *honey* excluders, as well, and refuses to use them. I agree with him, but I don't like having to harvest supers with brood in them—the frames that contain larvae should not be processed along with honey, and yet they present a number of cleanup problems.

I have worked out a compromise with the bees. Over the years, I have discovered that only a quarter of the hives have queens who go up into the supers to lay eggs, but there is no way to tell ahead of time which ones they are. So I put the first supers on all of my hives without a queen excluder, which is what I have been doing this past week. In two weeks' time I'll return to each hive with queen excluders and more supers. The bees in most of the hives will have stuffed the super with light-golden blackberry honey, a sight to gladden a beekeeper's heart. The filled frames of honey present a barrier to the queen, and in all likelihood she will not go up into any other supers I put on the hive, so on those I'll put two more empty supers, giving the entire hive a total of three, one filled and two still to be filled. But in some hives the center frames will contain eggs and young brood, indicators that the queen will continue to lay eggs in any other supers I put on her hive unless I confine her to the hive bodies below.

When I find such a super, I smoke the bees down as thoroughly as posible. The smoke usually sends the queen scuttling downstairs, too. Then I set the super to one side, carefully check each frame to see if the queen remains on any one them. If I find her, I gently carry her back to the hive and with my finger nudge her down among the frames in the hive body. Then I put a queen excluder in place on the second hive body and put an empty super on top of it. Above this, I put the super with the frames of brood. The worker bees will not abandon their brood. They will come up through the queen excluder to tend the developing bees, and in the process of passing through the excluder they will have become accustomed to it. They will store honey in the empty honeycomb in the first super and in the empty cells from which the brood have emerged just as readily as if there was no queen excluder there at all. But I must leave the queen excluder in place all season. Once a queen has proved herself to be the sort who likes to go up in supers to lay eggs, she seems determined to do so no matter how many supers of capped honey are between her and the empty combs. By the time I harvest the honey, all the brood in the frames above the excluder will have developed into adult bees and flown away.

I finished putting out the supers by midafternoon, and drove home. The day was hot, and I was wearing as little as possible under my bee suit. I indulged myself in a bit of beekeepers' air conditioning, angling my arm out the truck window so that a cooling current of air could find its way into the loose sleeve of the bee suit and funnel down across my hot, sweating body. I was so simplemindedly enjoying my own evaporative cooling that I did not at first notice the rotating red light of a patrol car behind me. I pulled over, and a young state trooper came running up alongside me.

"Your pickup's on FIRE, Ma'am!" he shouted.

I got out and walked to the back of the pickup with him to show him the bee smoker in its metal carrying case. The fuel inside was almost gone, but a few plumes of smoke still wisped from it. I explained about the bee smoker, and the trooper recognized my garb as the sort of thing a beekeeper wears. He blushed bright crimson and apologized for stopping me. I reassured him he was doing a proper job, and we parted friends.

When I got near home, my neighbor stepped out in the road and flagged me down. Surely *he* couldn't think my truck was on fire. He didn't, but he was excited, too. A swarm of bees had landed near his mailbox. Could I do something?

I walked over to the mailbox with him and looked. A group of bees, twenty-five thousand or more, had alighted on the thin branch of a slender bush, and their weight had dragged the branch to the ground. I retrieved his mail for him, told him the bees would probably spend the night there and I'd come up in the morning to remove them for him. I tried to tell him that bees were in a benign mood when they swarm—they have no home to defend so they seldom sting—but he'd rest easier, he said, if I'd come get them this evening.

Swarms are not much use to a beekeeper; they are headed up by an old queen, and to be of any value once they are hived she should be killed and replaced. Even then, because they are starting out on empty comb, the colony seldom produces any honey the first season a beekeeper hives them. Also, they may carry disease. But beekeepers are continually called in the late spring and early summer to pick up swarms, and usually we oblige. Because most people are rather afraid of bees, gathering up a swarm is an easy way to earn the gratitude of the person on whose property they have chosen to alight. But the real reason we pick them up is that it is so much fun to do so.

This year has been a famous one for swarming. All of us who keep bees are getting telephone calls about them.

A friend of mine, a banker who also keeps bees, has had more than his share. People even come into his office in town to report them to him. When I saw him the other day, he told me he had just been out to pick up a swarm that had wrapped itself around the trunk of a young tree. There was no way to shake them into the hive he brought to put them in, nor any easy way to entice them into it. So he put the hive on the ground, got back into his pickup and began to ram the tree with its front bumper until he had jostled them loose.

My neighbor's swarm would be easier to hive and I promised him I would be back directly.

IV

THE BEEKEEPER'S SUMMER

The Beekeeper's Summer

A swarm of bees in May
Is worth a cow and a bottle of hay.
A swarm of bees in July
Is not worth a fly.
"A Reformed Commonwealth of Bees,"
1655

A farmer today would scorn a deal offering him a swarm of bees, even in May, for a cow and a bottle (bundle) of hay. Cows at sixty cents a pound and hay at a dollar fifty a bale are pricey. And a package of bees (the equivalent to a swarm, but better because it contains a new queen) is not. One can be had for a mere twenty-five dollars from a bee breeder. But the basic truth of the rhyme still holds after three centuries. Swarms must be hived in May to be of any value at all and a swarm of bees in July, or even a package of them from a bee breeder, is indeed not worth a fly.

Early, or May, swarms have a chance of building up and proving useful to a beekeeper. A swarm of bees in July, when the flow of nectar from flowers has dried up, could not even keep themselves alive, let alone produce any extra honey to be harvested. Indeed, they would have to be fed lest they starve to death.

In the days of the rhyme, there were no bee breeders who sold tidy packages of bees, and picking up a swarm was one of the best ways to acquire them. Fashions in beekeeping change, as they do in everything else, and a hundred years ago in this country, when colonies died out mysteriously from moths and disease, the apiarist considered himself lucky to have hives which would swarm because it meant that he could replenish his dwindling stock of bees.

Bee breeders nowadays try to develop strains of bees that do not swarm, because contemporary fashion prefers them not to. Although the impulse to swarm may be written in the genes and therefore changed by selective breeding, the exact trigger mechanism of swarming is poorly understood. No one knows what makes one colony swarm and another not swarm. In general, crowded conditions within the hive seem to make bees more likely to swarm. And older queens are supposed to swarm more often than younger ones. However, anyone who has kept bees has had the experience of watching a swarm emerge from a recently divided colony with a brand-new queen in a hive with mostly empty comb. Beekeepers' generalities are flaunted by bees.

In some hives, the colony will swarm repeatedly during the same season. The first swarm, usually the one headed by the original queen, is termed the "primary swarm." The bees may have raised many new queens in swarm cells, and, instead of killing one another, the queens may fly out with decreasing numbers of the original population, in tiny secondary and tertiary swarms—a process that continues until the hive is nearly depleted.

It is exciting to see a swarm. The group of bees flies along in a stream, often low to the ground, and to someone attuned to their ways they seem to fill the air with electricity, giving a beekeeper a sense of elation—the human counterpart of the springlike, optimistic, burgeoning state the bees are in.

A number of my older Ozark friends can remember the days when bees flying overhead in a swarm were "tanged" to the ground. For centuries, beekeepers have been tanging—making a ringing noise by beating a metal spoon against a pot—under a swarm to bring it down. I've never tried it, and can't report any firsthand experience. But bees do not have a sense of hearing in the way we do. Gilbert White, the eighteenth-century English naturalist, established this fact to his own satisfaction by roaring at them through a speaking trumpet held near their hive. He reported they failed to show "the least sensibility or resentment." Modern authorities say bees are virtually deaf to airborne sound but moderately sensitive to sound waves that travel through the ground or through other solid objects, which they can detect through sense organs on their feet. These experts say tanging does not work, and any swarms that descend while being tanged were going to come down anyway.

Before leaving the parent colony, the bees stuff themselves with honey, so they have no need for food for several days. After they have swarmed, the bees lose all memory of their old hive and its location. (It would be interesting to know how this "unlearning" takes place.) They light temporarily in some exposed spot—a tree limb, the side of a building, an automobile hood—while scout bees search out new quarters. In the springtime newspapers often carry stories and pictures of swarms of bees which have caused some excitement by alighting in an inconvenient location, on some downtown building, perhaps.

Passersby are seldom aware the bees are not cross, will not sting them, and, if left to their own devices, will leave in a day or two. Beekeepers like to show off as much as anyone, and when they are called in to help with those urban swarms they enjoy giving the press an animal-taming act by picking up the swarm without a veil and with their bare hands. It is easy enough to do. The bees do not develop a defensive sense of place about their temporary hanging-out spot, and so they are seldom aggressive unless they are attacked by humans who try to squash them or spray them with insecticides. They will cluster around their queen in this temporary location until the scout bees find them suitable new quarters, a process that may take several hours or may stretch out to several days.

We know about two forms of honeybee communication. One is chemical: information about food sources and the well-being of the queen and colony is exchanged as bees continually feed one another with nectar or honey that has been chemically tagged by the bees while processing it. The other is tactile: bees can tell other bees about things such as food or the location of a new home by patterned motions. These elaborate movements, which amount to a highly stylized map of landmarks, direction and the sun's position, are called the "bee dance," and have been described in *The Dancing Bees* by Karl von Frisch, who discovered this behavior. Entomologists are currently questioning some of his findings, but von Frisch claims that the directions given in the dances are so precise and accurate that once they have been deciphered researchers can go to the place indicated by the dance and watch tagged bees arrive.

It is an elegant form of communication, more sophisticated than that used by other social insects—ants, for instance, recruit nest mates too, but they merely drag or carry their fellow workers to the spot where they want them to be.

Different scout bees may find different locations for the swarm and return to dance about their finds. Eventually an agreement is reached, rather like the arrival of the Sense of the Meeting among Quakers, and all the bees in the swarm fly off to their new home.

The bees up by my neighbor's mailbox were clustering there and receiving scout bees, who were investigating all the snug spots they could find, including various hollow trees and spaces under the siding of the abandoned cabin in back of my mailbox. If I had left them alone they would have selected one of those places and flown away to bother my neighbor no more.

But I interrupted the process. I brought with me one covered single hive body stapled to a bottom board. I had taped its ventilation hole but had left open the entrance, although I had also brought along a screen for it. Inside, I had put a feeder and eight frames of worked comb which, to a bee, would be fragrant with the odor of other bees. I also brought along a white sheet.

My neighbor and his family gathered at a distance to watch. A good portion of the swarm was on the ground because the weight of those thousands of bees had dragged down the small branch. I placed the hive on the ground, near the bees, with its entrance toward them, and spread the sheet in front of it. They would have to walk over the sheet to get into the hive, and the white background would make it easier for me to see the queen. I had to make sure she was in the hive, or the swarm would not stay in it.

It took a few minutes for the bees to become aware of the hive with its welcoming scent of comb on which other bees had lived. The first bees to discover it stationed themselves at the entrance, elevated their abdomens and pointed them outward to expose their scent-producing Nassanoff glands. They fanned their wings to send out an odor trail from these glands for other bees in the swarm to follow. Tired of camping out, the entire swarm started walking slowly across the white sheet and into the hive, in a purposeful and what can only be described as a cheerful manner. I sat astraddle the hive watching for the queen.

The thousands of bees, even though they were marching along at a steady clip, took quite a long time to go into the hive. My neighbors at first found it comical that I had somehow induced the bees to walk into the hive on their own, but soon they grew bored. They had expected something more lively. They had assumed I would *do* something, not just sit there on a hive and watch the bee parade. My neighbors had long since left when I finally spotted the queen stepping briskly across the white sheet. She was a beautiful golden yellow with three stylish black stripes, marking her as an Italian. She disappeared into the hive. I got up and stretched. It took a bit longer for the rest of the bees to follow her in. Then I shoved the entrance screen into place and stapled it. I put the hive into the pickup and drove home, where I unloaded it out in back by the woodlot. Tomorrow morning, I'll fill the feeder full of sugar syrup to give these bees a start. In a week's time, I'll medicate them and check to see what kind of an egg-laying pattern the queen has made, before I decide whether to let her live or replace her. I hope she is still a good queen—I shouldn't like to have to kill such a beauty.

A friend who keeps bees stopped over at day's end for a drink. We sat up on the deck of the barn loft and watched the sun go down. I told him about the swarm and the beautiful queen. We talked about how odd and special it is to develop an affection for bugs. He laughed and proposed a toast.

"To the queen!"

"Long may she live!" I responded.

And then we watched the sun disappear behind the trees.

The banker and I are keeping quiet about the bees in the bank parking lot. They are flying around in a desultory way; we both know they are spiritless and won't sting anyone, but there is no need for us to let on that we know those bees. They are stragglers from a demonstration our local beekeeping organization held last week to show beginners how to hive a package of bees. We had asked a retired commercial beekeeper to do the demonstration, and he kindly obliged. He has been a great help to our local organization, giving us all the benefit of his years of experience as the owner of a large honey company in one of the western states. Western desert beekeeping is unique, but like big commercial operators in other parts of the country, this beekeeper often moved his colonies around in order to capture different blooms. These colonies were moved within the state only, but in a typical commercial operation in other parts of the country beekeepers will move colonies from, say, Florida, where they have worked citrus flowers, to Ohio for spring dandelion blossoms and then on to the Dakotas, which is one of the best honey-producing sections of the country because of its long summer days and ample bloom of clover, alfalfa and native plants.

These commercial honey companies run thousands of hives, which are placed on pallets so they can be mechanically hoisted for moving. This involves a large investment: fork lifts, boom loaders, tractor-trailer rigs, honey-extracting plants with sophisticated equipment and a large labor force. Honey in other countries can today be produced more cheaply than it can in the United States, and most of the honey on grocery store shelves is of foreign origin. For a time, a federal price-support program kept the big American honey producers in nervous but marginally profitable financial health, but the price-support levels have been cut in recent years, and gloom has settled in a big black cloud over them as it has over other segments of the farming community. Some of them are cutting back and selling out.

I am the only beekeeper in our local organization who fits the commercial definition by government standards: a producer who has at least three hundred hives of bees and makes her primary living from them. Most of the other members have a few hives, and some of them even sell their honey. The banker, for instance, uses his bees to make comb honey, which he packs attractively and which I peddle for him along with my extracted honey in jars. However, none of us have had the broad experience of the retired commercial beekeeper. He is a quiet, softspoken man, generous with his time and knowledge and happy to help newcomers get started with bees. On the Sunday scheduled for our demonstration, he came early to set up his equipment. We had ordered a three-pound package of bees from a southern bee breeder for him. (Package bees can be bought in two-, three- or five-pound sizes.) The bees come in secure cages made of screenwire stretched across a wooden frame, allowing them to have air but preventing them from escaping in the post office—although postal workers are never as sure about that as beekeepers are. Each package contains the specified poundage of bees, a queen in her own small cage and sugar syrup in a can for food during shipping.

We had brought along what a beginner would use for hiving a package of bees: a covered hive body on a bottom board, a feeder and a full complement of frames containing foundation only. It is easier to hive a package of bees on combs that have been fully worked, because the bees are more likely to want to stay on them, but a beginner's equipment is usually new, and includes new frames with foundation.

Package bees are disoriented from their trip through the mails;

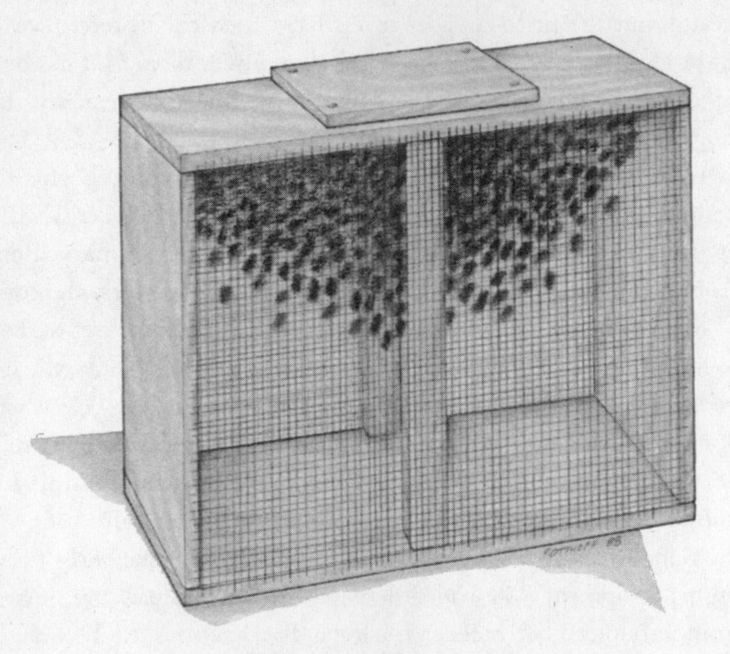

Bees in mailing cage

they have often been scooped up from different colonies with different chemical identities, and the queen they are given is new to them. They are out of sorts when they arrive, and may fly away from the hive into which they are placed. Package bees are best installed as late as possible during the day because, as darkness comes, they will be less inclined to fly; if they have one night to get used to their new quarters, they will be more likely to remain in them.

Beekeeping association members and others curious about bees began to gather in the bank parking lot, and by the time the demonstration was about to start we had nearly fifty people standing in a respectful circle around our retired beekeeper. He first stuffed up the entrance of the hive with some grass, to make it harder for the bees to use; then he shook the mailing cage to force most of them to drop to the bottom. He removed the wooden cover deftly, and before too many bees had escaped he pulled out the queen and replaced the cage cover. He removed the cork over the sugar-plug end of it, and wedged the cage down among the frames. Then, through the screen, he sprinkled the bees inside the big mailing cage with sugar syrup. They began sucking it up and cleaning one another off, and while they were all occupied with this, he again took off the wooden cover and dumped the bees on to the frames of the open hive. Most of them obligingly crawled down into the frames, but others flew up in confusion and disorder. He had moved quickly and surely and had not angered the bees, however, so they bothered no one. After he had closed up the hive, people gathered closer to ask questions—ignoring the bees circling in the air. It was nearly dark by the time their questions were answered, and many of the loose bees had found their way into the hive. We put it into the pickup of the man who had paid for the bees.

Back at his place, the bees should release the queen within a few days. Once the beekeeper has taken them home, he will continue to feed them steadily with sugar syrup until they have drawn out the foundation wax in the frames, built honeycomb on them and filled each hexagonal cell. After the first hive body is nearly filled, he will pull out the extra frame and add a second hive body. He will put the one drawn comb taken from the bottom hive body in the middle of the nine frames of foundation in the new upper hive body, thus temporarily giving the second hive body ten frames. The one drawn comb frame will help lure the bees upstairs to work on the new foundation. He will continue to feed sugar syrup, and after the frames in the second hive body are also drawn and filled he will remove the extra frame and thereby reduce the number of frames to the usual nine. He will space them carefully to respect the bee space.

There is no way to tell what the weather will be in the summer ahead, nor when the rains will come, but in an ordinary year these bees will do well to draw out and build comb on the frames of foundation and fill them with enough honey to keep themselves through the winter. In an exceptional year, package bees started on foundation may produce some extra honey that a beekeeper can harvest, but he should not expect it. He should put off any plans to take honey from them until the following year.

Our local beekeeping organization, an affiliate of the state association, was started because we thought it would be good to get together and talk bees. Demonstrations like the one we had last Sunday have also given us the satisfaction of helping a number of new beekeepers. For me, whose livelihood depends on bees, there has been another benefit. Our group is a force worth listening to in dealing with what is always a problem for beekeepers—the loss of bees to pesticides.

Insecticides are not heavily used by farmers in the Ozarks, who grow few crops but raise cattle and hogs instead. In some parts of the country the liberal use of agricultural chemicals has made beekeeping all but impossible. Insecticides kill the bees, and herbicides kill the wild plants bees use to make honey. The bootheel of Missouri is an agricultural area, where cotton, soybeans and sunflowers are grown in big fields. Potentially each of these crops is a good honey source, and the only really big commercial beekeepers in the state are based there to take advantage of them. But a few years ago a friend of mine in the bootheel, a man who was running 7,000 hives, lost ninety percent of his bees to insecticides and was forced out of business.

Aside from the economic loss, it is a discouraging sight for a beekeeper to find masses of dead bees twitching in their death throes in front of a hive he has carefully nurtured and brought along all spring. Some agricultural pesticides act so rapidly that the bees die in the field, but with others the bees struggle back and die in convulsions in their hive, where, as long as workers remain alive inside, they are carried out and piled in growing heaps in front of the entrance. The honey and pollen may be contaminated, and the work force so depleted that the hive will probably die out, even if the workers have not brought home insecticides so potent that they will kill on contact all the larvae and developing brood, although that sometimes happens, too.

The worst of these losses can be easily prevented if a beekeeper knows a day ahead of time when insecticides are going to be used. The night before, he can tape shut any holes in the hives and screen the entrances to keep the bees from flying out. If he is able to confine his bees to their hives for twenty-four hours after spraying, the worst of the insecticide residue will have broken down and bee deaths will be kept to a minimum.

Farmers and agricultural pesticide applicators are not evil men. They do not deliberately try to kill bees, but they forget that bees are insects and insecticides that kill "bad" insects will also kill "good" ones like honeybees, ladybugs and butterflies. So they need to be made aware of any colonies of bees kept in an area likely to be sprayed. Since aerial sprayers seldom think about us, it is up to beekeepers to find out what the schedule is on any farm near their bees. Here in the Ozarks, even without the extensive use of agricultural chemicals, beekeepers lose bees during fruit bloom, when orchardists are spraying their trees, during spring alfalfa bloom, when farmers hire aerial insecticide sprayers to treat for alfalfa weevils, in summer when the same alfalfa is sprayed for grasshoppers and later still for armyworms.

After I lost forty hives of bees to fall armyworm spraying, I worked through the beekeeping organization to see what we could do to prevent such losses in the future. We invited the extension agent to speak to our beekeepers' group about local spray schedules. He was helpful, and was also surprised and impressed that there was a large and responsible group of people in the area, a group that had every reason to be concerned about the use of insecticides. We became a part of his constituency, as it were, and he began including a warning about the danger of pesticides to bees in his press releases and spray recommendations. I also wrote news releases to help increase awareness about bees and insecticides from a beekeeper's point of view, and the local newspapers and radio stations were glad to run them. The extension agent gave me the names and addresses of all the commercial pesticide applicators active in the area, and I wrote asking them to call me before they applied insecticides aerially so that I could notify the beekeepers within the area. The beekeepers and I mapped the location of every beekeeper we knew within a five-county area and made a file of their telephone numbers, so that I could call them if I heard about spraying. Unfortunately there was no cooperation from the commercial aerial applicators even though they are legally liable for any damage they do—which includes colonies of bees killed—but I did end up with friendly help from the foremen at the big ranches who hired them. These are local men, neighbors, as the aerial sprayers are not, and a jar of honey made them remember me. Now they call me the day before the aerial sprayer is due to treat their alfalfa, and I notify any of the beekeepers who have hives located within two miles of the spray area. The beekeepers then can take steps to protect or move their hives.

The system is not perfect. We all still lose some hives to insecticide poisoning, especially when people spray their home gardens, but we lose fewer than we did before we were working together, and in these days of chemical agriculture that is no small success.

To hear people talk down at the café you'd think we were being invaded by hostile aliens from a Grade B science-fiction movie. A new brood of thirteen-year cicadas, estimated to be in the millions, has crawled up out of the ground here in Missouri, as well as in portions of Illinois, Arkansas, Oklahoma and Kansas. Its members are trying their wings, singing like crazy and mating in a very public way. In a few weeks they will die off and be gone, but for now they are quite the topic of conversation.

There are annual cicadas here and all over the United States, and their buzzy song is a part of the sound of late summer in the country, like those of katydids and whippoorwills, but in the eastern half of the country, at long intervals, periodic in early summer, cicadas emerge from the ground where in their juvenile form they have been feeding on the sap of tree roots. In the northern part of their range they appear every seventeen years; in the southern part, it takes thirteen years for a new brood to emerge.

These periodic cicadas are not so big as the annual variety but they are large as bugs go, broad, sturdy and two inches long. They have midnight blue, almost black, bodies and huge showy wings that are delicately veined and edged in brilliant orange, and look as if they are made of isinglass. Their legs are orange, too. Their big, bulging eyes are deep red, and each is centered with a meaningful-looking dark-brown dot. They are beautiful, spectacular bugs, and their appearance is exuberant and exciting, a marvel, a celebration. Their presence is a reminder that there are other life cycles than ours, other rhythms of living than the human one.

When they first come out of the ground they are still wearing their golden-brown nymphal skins. After crawling out of these wingless skins, which split neatly down the back, the handsome adults sit quietly and visibly on tree trunks, tentatively trying their new wings. Then they enjoy a few weeks of courtship and mating. During those weeks the males sing, the paired couple reproduce and then they both die.

Entomologists speculate that the periodic cicada's infrequent but abundant aboveground appearance is a survival strategy. Predators cannot thrive on a dinner that shows up so seldom, and the cicadas' sheer numbers guarantee that even though individuals are easy to catch many will survive. Some also claim that the cicadas' peculiar, burning, raspy song, produced by a pair of ribbed membranes at the sides of the male's abdomen, is irritating to predators. That seems fanciful, but perhaps it is true; people are uncommonly cross about cicadas and complain their song is nerve-racking. Here it is certainly constant and pervasive. The cicadas have three things to say: one is a steady, insistent, buzzy trill: *zs-zs-zs-zs-zs.* It is a background to a more varied *kee-o-keeeee-o-kee-o* that punctuates the steady drone. When picked up and held, the cicadas emit a sharp *bzz-t-byzzt* that sounds troubled and probably is.

Once the insects have paired, they mate openly on tree trunks and branches without regard for the fact they can easily be picked off and eaten by predators or squashed and dusted with poison by humans. But most survive, and the female is able to lay upward of five hundred eggs in slits she makes in twigs. The lives of the adults are short, but in eight weeks the cicada nymphs hatch and burrow down into the ground to reach the tree roots on which they will feed and grow slowly for the next thirteen years.

The periodic cicadas do not kill trees in their feeding, and at no point do they hurt garden vegetables or healthy flowering plants. They are bugs of such innocence and beauty and specialness that their appearance, one would think, should be regarded with interest and appreciation like that of a comet or a rare bird. But it is not.

Down at the post office, the talk among people waiting with me for the mail was of how bad the cicadas are and how much worse they will get as they become more numerous. Over at the café the morning coffee crowd was discussing how to do them in. Squashing them between thumb and forefinger was held to be effective but unaesthetic. A stick was recommended for bashing them. Since the adult cicadas do not chew leaves, only contact poisons will kill them, and the effectiveness of various kinds was being hotly debated. Some of the official types have been recommending dusting them with serious insecticides, but one of the coffee drinkers remembered the local state forester had said that this could cause increased levels of pesticides in the groundwater supply. The coffee cups were drained without reaching an agreement on methods, but the general opinion was that these bugs are annoying, lascivious, untidy, unruly—in short, a nuisance.

Back in my woods, where I have begun cutting the winter's firewood early each morning, the cicadas' song filled my head, seemed to reverberate inside it. Cicadas, the sun catching their wings and reflecting rainbows, lined every tree trunk, every branch. One lighted on my shoulder. His broad face with its big red eyes was inches from mine.

Kee-o-keeeee-o-kee-o he sang zestfully, right into my ear. He sounded pleased with himself; I know I was mightily pleased with him.

Farmers began their first cutting of hay weeks ago. I should have taken my brush cutter and trimmed around the beehives then, too: the grass and brush has grown tall around them, making it hard for me to work the hives, and it has also made it difficult for the bees to fly in and out. I keep scraps of rolled roofing in front of the entrances to help keep them clear, but the grass has become overgrown and sprawled across the roofing, so some entrances are blocked. However, I have been out on the road making my June honey sales trips. I've been to Boston, New York, St. Louis, Kansas City, Tulsa, Oklahoma City and Dallas. Over the years my bees must have become more skilled, for I have more honey to sell, and with the competition from foreign honey there are more of us scrambling for a smaller slice of the market. I should be setting up accounts in other cities as well. But I'm already spending more of my time peddling honey than I am keeping bees, and I should prefer it the other way around. Just taking care of my established accounts throws off my schedule of bee work. Out on the interstates, I was seized with guilt about the bees, so this past week after I have finished cutting wood each morning I've been working hard in the beeyards. I have been trimming, and taking more supers to the hives. They are needed.

The nectar flow is at its strongest now. Most hives are ready for at least two more supers. Individual variations among the colonies of bees, which a few months ago were exactly equal, are now apparent: there are some hives, with hardworking populations of bees, that have five supers on them and will need more before the end of the nectar flow; there are others with only one super, and the bees may not even fill that one. Hot weather is ahead, and it is kind to help bees ventilate their hives. With those that have four or more supers, I offset the top super slightly, leaving a crack between it and the one underneath. The space allows the bees to set up an air current from the hive entrance to the top, helped on its way by their fanning wings, and so lets them both cool the hive and evaporate the nectar they are storing in the supers. Some beekeepers achieve the same effect by propping open the cover slightly. Either method is satisfactory and helpful to the bees.

Colony populations are at their peak now, and the bees in the best hives are working in a dedicated way, flying out in streams and coming back to their hives heavy with nectar. The Ladino clover, *Trifolium repens giganteum,* is in heavy bloom in farmers' pastures and yellow-blossomed sweet clover is almost through blooming along the roadsides, with the white-blossomed variety just coming on. Those tallest of clovers—*Melilotus officinalis* and *M. alba,* respectively—grow from five to eight feet high. A hundred years ago, they bloomed thickly from the Rocky Mountains to the Mississippi River, and their deep roots brought nutriments up to the surface of poor but limy soils. Today, beekeepers still think of sweet clover as the best of all possible bee plants, and have been known to carry sacks of the seed to sow broadcast wherever there is a likely spot. Other kinds of farmers like it less well, and seldom plant it. In the first place, sweet clovers are biennial, and with their rangy, weedy growth they do not make as good fodder as do the lower-growing clovers such as Ladino. So sweet clover today grows mostly in waste places such as highway rights-of-way where the bees make use of it until the highway crews come along and mow it, a sight that always makes beekeepers sad.

This week, my bees are also working persimmon flowers. Wild persimmons, *Diospyros virginiana,* are skinnier trees in the Ozarks than they are in other parts of their range, but they still grow tall here. Their white, bell-shaped flowers—male and female borne on separate trees—are too high for people to see, but bees find them. The flowers are such a good source of nectar that the bees are often out in the persimmon grove by the woodlot before dawn. Some bees even spend the night on the blossoms; they are so reluctant to stop working them at day's end that darkness catches them there, and they cannot fly home.

All week, as I have worked with the hives trimming grass and adding supers, the bees have been so busy in their foraging flights that they barely noticed me, even when I trimmed directly in front of their hives. In the middle of a sunny day when there is a strong nectar flow it is almost impossible to make bees cross, and easy to work among them without gloves, veil or protective clothing. I have kept my smoker burning as a precaution, but I've seldom used it: even in hives that have developed into touchy cross strains, the bees are impossible to provoke to anger under these conditions. Yesterday, however, was different. Even if I had not heard the weather forecast on the radio in the morning, I should have known we were due for a change in the weather by the bees' behavior. The meteorologists and the bees both told me a front was moving in from the Rockies which would bring rain by evening, and although the day stayed sunny, the air was oppressive and the bees had hair-trigger tempers.

I use a heavy-duty gasoline-powered brush cutter to trim around the hives. It is fast, and the bees, of course, do not mind its noise, though I do. To me it sounds like an outboard engine on a small boat. With it, I can trim near the hive entrances, but I must finish the job on hands and knees with a pair of clippers. It was my position at their entrances, not the noisy brush cutter, that irritated the bees the most. Earlier in the week, when the weather promised fair, they had ignored me, flown around me, when I trimmed at their entrances with the clippers. But yesterday when I crouched there, the guard bees bristled. They are older workers with full venom sacs, ready to sting, and they recruited helpers among their sisters to come out and try to drive me away as I inched along the front of the rows of hives. I smoked them heavily to quiet them. If I had not and if they had stung, every bee who left her stinger in my bee suit would have hung a chemical label on me: "enemy" her pheromone tag would have said. That message would have been read by bees from hives I had not yet come to, and made the bees attack before I approached them. Within a few hours, even with my dull human senses, I should have been able to smell the acrid alarm pheromone, the scent of angry bees.

I did not have any real trouble with the bees yesterday, because I used smoke to gentle them and because I have worked with bees so often that I am used to them and their moods. But I knew they were touchy, and if I had jarred the hives accidentally or disturbed them severely their anger would have erupted. I don't like working with them when they are so clearly out of sorts. I can't think it is kind to them to do so. If I had fewer hives and a looser schedule

I should have put off working with them entirely yesterday. One of the luxuries of keeping just a few hives of bees is that one can work with them on fine days only—reason enough, sales trips to Dallas aside, to own three or thirty, not three hundred hives of bees.

I do not like to irritate bees, and by day's end yesterday I was cross, too. Perhaps the change in barometric pressure and the threat of the storm affected my mood as much as it did the bees'. A magnificent electric storm and heavy rains are keeping both the bees and me in our respective houses today. The storm is clearing the air of oppressive heaviness, and tomorrow the bees' mood will be better. So will mine.

It is the Fourth of July, and the local VFW roasted a pig at the park in town today before the evening's fireworks. My friends among the veterans invited me to come, but I was too busy. I have been working out in the beeyards, putting on the last supers of the season. It has become my own private holiday tradition over the years to get the final super on the hives before the fireworks go off.

The dates of the honey season are different in different parts of the country. They depend on local weather conditions and flower bloom, but here in the Ozarks the nectar flow is usually over by mid-July. Hot and dry weather in July worries farmers who graze cattle, because their pastures dry up; if it rains, pastures will green but the sumac nectar flow will cease, so I am the one local farmer who prefers midsummer droughts. Heat and dry weather in early July make sumac blossoms secrete nectar, which the bees transform into a wonderfully rich-tasting dark-red honey. There is no part of the country in which some species of sumac, all members of the genus *Rhus,* does not grow, and wherever the weedy trees are tolerated bees make honey from their flowers. It is one of the most important honey sources in New England.

Here in the Ozarks, one of the first plants bees work in the springtime is *Rhus aromatica*—fragrant sumac. The last significant

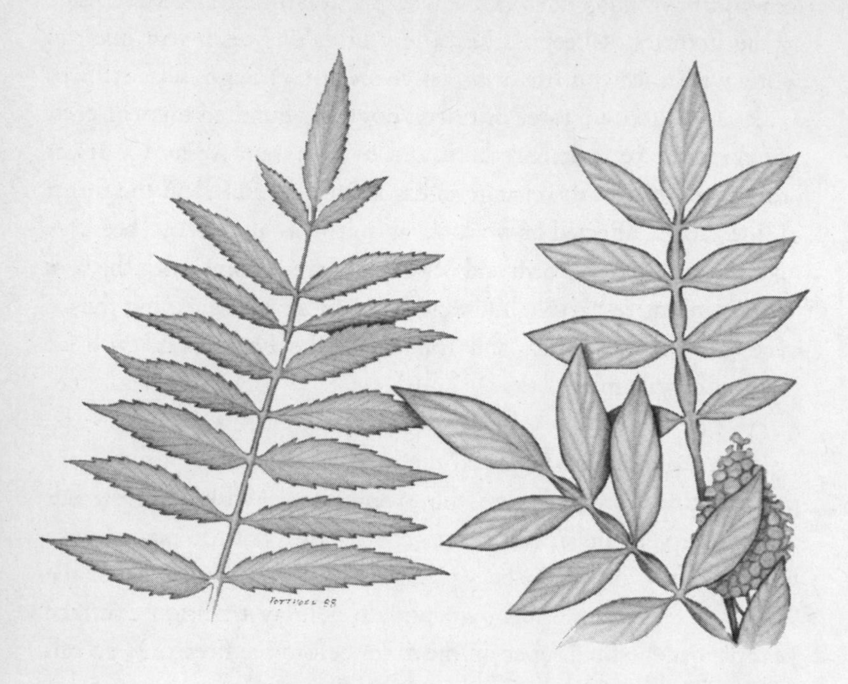

R. glabra *(left) and* R. copallina *(right)*
(smooth sumac and shining sumac)

nectar flow of summer comes from its cousins, *R. glabra* (smooth sumac) and *R. copallina* (shining sumac). Smooth sumac is the first of the pair to bloom. It is a small shrubby tree seldom over ten feet high and is one of the first invaders of abandoned fields. Its leaves, alternate, borne compound, turn brilliant red early in the autumn, and its creamy-white blossoms are borne in clusters followed by the crimson globe-shaped berries that eastern bluebirds find tasty and sustaining in winter. In the hot dry weather, which stimulates nectar secretion, the white blossoms drip nectar and are covered with bees, who lap it up. Later, if the weather holds, the bees will transfer their attention to *R. copallina,* a sumac similar in growth, flowers and fruit. It can be easily distinguished from *R. glabra,* however, because the leafstalk is edged with small green wings, which give it another common name, winged sumac. Together, the two species provide weeks of bloom, and during a hot droughty summer, sumac is the important source of a delicious honey that I will harvest to blend with the rest of the crop. But, like other beekeepers, I have learned that customers are alarmed to learn that bees make honey from sumac flowers.

"Sumac!" says one, screwing up her face to indicate her distaste. "Isn't that poisonous?" I had heard the question a number of times.

Poison sumac, *R. vernix,* grows along the eastern coast and in boggy places throughout the southeast, not in this part of the country. However, people have had such bad experiences with the maddeningly itchy rash it causes they are wary of any plant with "sumac" as part of its name. Earlier in the season, bees here *do* work another member of the same genus—*R. toxicodendron,* poison ivy, which causes a similar rash. But the honey that bees make from it does not harm human beings. In fact, there is only one plant in this country from which bees produce honey generally poisonous to people. It is the mountain laurel *(Kalmia latifolia),* the evergreen shrub that frosts eastern mountains with its white flowers in the springtime. Mountain laurel contains a poison called andromedotoxin and its effect has long been known. Writing in an 1875 issue of *Gleanings in Bee Culture,* a surgeon who had been with the Confederate army wrote:

> Wherever the Mountain Laurel grows the bees are very fond of it. . . . It is dangerous for any one unable to detect the taste to eat the honey. It has a highly poisonous effect, being an extremely distressing narcotic. . . . During the war I had many opportunities of witnessing its effects and on one occasion, personal experience gave me the right to say that I knew something about it. . . . Some time after eating a queerish sensation of tingling all over,

> indistinct vision, caused by dilation of the pupils, with an empty dizzy feeling about the head and a horrible nausea which would not relieve itself by vomiting. . . . The first case or two that I saw were entirely overpowered by it, and their appearance was exactly as if they were dead drunk. . . . The enervation of all the voluntary muscles was completely destroyed. The usual remedies for narcotics partially restored them in a few hours, but the effects did not completely wear off for two or three days. . . .

There are, to be sure, variations in what triggers an allergic response in people, and green, or unripe, honey from a variety of plants (yellow jasmine and rhododendron, for example) can cause reaction in susceptible humans. In other parts of the world, there are other poisonous honeys. Xenophon describes soldiers in his army who were driven mad by eating honey. Strabo and Pliny wrote similarly of a honey called "goat's death," which modern writers believe to be made from *Rhododendron ponticum,* a little-cultivated species. There are also intriguing reports of a Brazilian plant, *Serjonia lethalis,* which produces a honey so poisonous that Indians used it for tipping their arrows and killing fish.

But in this country only mountain laurel gives nectar that humans in general should avoid. Bees are not so lucky. There are a number of plant nectars that can poison them. The most important ones in this country are California buckeye *(Aesculus californica)*; black nightshade *(Solanum nigrum)*; death camas *(Zygadenus venenosus)*; dodder *(Cuscuta spp.)*; leatherwood *(Cyrilla racemiflora)*; locoweeds *(Astragalus spp.)*; mountain laurel *(Kamia latifolia)*; seaside arrowgrass *(Triglochin maritima)*; whorled milkweed *(Asclepias subverticillata)*; and western false hellebore *(Veratrum californicum).* The symptoms of plant poisoning in bees are similar, in some respects, to those of insecticide poisoning: bees may die in heaps outside their hives. But there are other symptoms as well. Newly emerged bees may have crumpled wings, or fail to shed the last pupal case from the abdomen. California buckeye poisoning causes bees to become black and shiny from loss of hair and may make them tremble. Affected queens produce eggs that do not hatch or larvae that die soon after hatching. The queens may also become incapable of laying or may lay only drone eggs. In cases of plant poisoning reported from Florida and Georgia, larvae in the cells turn blue.

I know all this at second hand, for I have never seen a case of plant poisoning in my hives, and my bees work no plants that produce honey hurtful to human beings. In a month or so, however, if droughty weather continues and absolutely nothing else is in bloom, the bees may be forced to gather nectar from bitterweed—*Helenium amarum tenuifolum*—which grows in pastures here as well as in much of the eastern half of the country. Another common name for bitterweed is sneezeweed. The crushed flowerheads have been used for snuff, which causes violent sneezing. Dairy cows grazing on the drooping rayed yellow flowerheads give bitter milk, and bees produce distasteful honey from its nectar. One year, I had a few combs of it in my harvest. The honey was beautiful to see—thick and creamy yellow, rather buttery in appearance—but it tasted like soap. To bees, it is as good as any other, and so I gave it back to them. Bitterweed is usually not a problem, because it is a meager nectar source and bees will not gather it if they can find anything else in bloom, even lawn plantain, *Plantago lanceolata*. ("Your bees must be starvin'," lamented a friend in town. "Why, they was workin' them little bitty stems in the lawn. Poor things. Just stems!") But I do not like to risk mingling bitterweed honey with the better-flavored honey from blackberries, clover, persimmons and sumac, so I always harvest my crop before the bitterweed blooms widely.

The fact that bees, in some few cases, make honey that is distasteful or even poisonous to humans points up a fact overlooked by those who do not live on familiar terms with "different bloods," as

C. S. Lewis calls them. Although living organisms are alike in that we are all animate bundles of carbon, hydrogen, oxygen, nitrogen, sulphur and phosphorus, we do things very differently. Insects and humans live very different lives, make different demands on the world and respond to it differently. Most people find honey tasty, but I am astounded when people call honey, a dense carbohydrate that is an excellent heat-producing fuel for bees, a "health food." I am astonished by the miraculous claims made for pollen, a high-protein food for young bees, in the human diet. I am amazed to see the high prices humans will pay for royal jelly, which is needed to transform a worker bee larva into a queen bee.

"Bugs is bugs," wrote Peter Eicher of Jackson Heights, New York, in response to an article I wrote about insects for *Time*.

Amen, Mr. Eicher. Amen.

And for those who think otherwise I should like to recommend a delightful story by Roald Dahl, "Royal Jelly." In it the bee-keeper-hero eats royal jelly and grows a covering of fine golden hairs.

The sun has been rising fiery red each morning for weeks now, and shining fiercely on dried-out fields. The temperature reaches a hundred degrees or even more by midday, and doesn't drop below it until the sun sets again.

It is so hot I don't feel like doing much in the middle of the day, and fortunately I don't need to. I have done everything I can to help the bees, and from now on until the honey harvest they are on their own. I have been spending the hot part of the day down at the river, where the water is still cool. Sometimes I take an inner tube and float from one river-access road to another—it makes me feel very entrepreneurial to see my bees along the river hard at work on the water-willow flowers, while I am playing in the stream. The water-willow leaves do look uncommonly willowlike, but *Justica americana* (the genus name honors James Justice, a Scottish gardener)

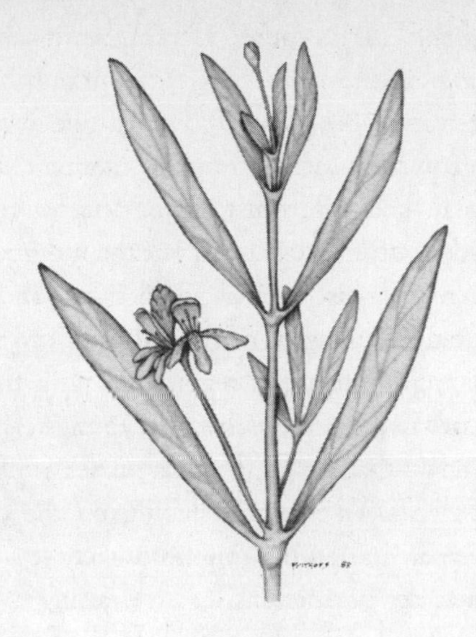

Water willow

is not a willow at all, but an acanthus. It is a low-growing, water's-edge plant of such modest habit that it usually goes unnoticed, even though it is one of the commonest plants there. Had it not been for the bees, I probably would never have had the pleasure of making the acquaintance of water willow. The flowers, which bloom from late May until October, are a good source of nectar; each one consists of a three-quarter-inch tube that opens into a notched upper and a three-lobed lower lip. They are a beautiful shade of lavender, and resemble tiny orchids. I keep pointing them out to friends when we float. Even those who have been on the river for years are seldom familiar with the flowers. I am grateful to the bees for showing me water willow, and I wonder what other beauties I fail to see for lack of suitable guides to point them out.

I have been thinking lately about how much I miss when I am as busy as I have been over the last months, because now, during the summer slack time, I have become the observer of several dramas. I am sure there are others I have never had time to see. Over the past few weeks, for instance, I have been watching a baby cowbird grow up in an indigo bunting nest out in back of the honey house. I first discovered the gangly cowbird two weeks ago, when he had already grown too demanding for the male indigo bunting who was trying to keep the noisy changeling in food. A female cowbird had, as they often do, pushed out the indigo bunting's eggs from the nest in which his mate had laid them, and laid her own eggs there. I had not been aware of it until I noticed an odd pairing of birds. A female goldfinch had teamed up with the male indigo bunting to help satisfy the greedy cowbird youngster. The baby bird is doing nicely now, thank you, but his nurses are being worn to a frazzle by his needs. I've been checking on the odd household every day. The female indigo bunting is nowhere to be seen. Perhaps she gave up when her partner insisted on raising a cowbird.

Then last week, while I was having my morning coffee, I discovered another bit of excitement on the woodpile behind the cabin. I am cutting firewood in earnest now; I spend a few hours each morning out in the woodlot with my chainsaw before the day gets hot, and I have been stacking the wood in back of the cabin. There, against the wood I had first cut, a spider had spun her web and laid eggs, which I had never noticed. Last week there were spiderlings as thick as stardust on the web.

They were so small, not even the size of a pinhead, that they must have just hatched. There were hundreds of them tumbling over one another in the web. I went back into the cabin for a magnifying glass, so that I could look at them more closely, but they were so tiny and I am such a poor arachnologist that I could not decide what kind they were—and their mother, who might have been easier to identify, was nowhere to be seen. During the day I checked on them several times. I could see them wandering out over the woodpile, but they must have returned at dark because the next morning I found them bunched again on the web, which was beginning to show signs of wear. There were not as many of them as on the previous day; some must have died during their initial exploration of the world. Or perhaps they found other places to congregate. The second afternoon, when I came back from the river, I found them all spread across the woodpile again. The following morning I found a cluster of perhaps half the original number on the torn web; by the time I had cut a truckload of wood and brought it back to stack they were gone. Had they been eaten? Were they foraging beyond the woodpile? In the evening a rain shower, welcome to humans, blew in from the southeast. At dawn just a few spiderlings were huddling on the web, which was badly ripped. Had the others drowned? I never saw them again. Those few disappeared during the day, and the next morning there were no spider babies to be seen anywhere. Perhaps they molted and moved on. The few scraps of web that remained fluttered in the light breeze.

Last evening, after it had cooled off a little, I walked back to the beehives by the woodlot. There are twelve hives there, their combined work force amounting to approximately 720,000 bees; every one of them must have been fanning nectar to evaporate moisture last night, because the hum of wings sounded like some immense machine in operation. My honey factory was in full operation. I held my hand in front of several of the entrances and ventilation holes, and could feel the moisture-laden air being pushed from the hives. For fifteen years now I have worked on such familiar terms with the bees that when I see them down at the river, or listen to them at night, I know exactly what they are doing. I now can understand them a bit, though not nearly as much as I thought I did the first year I worked with them. They have forced me to realize that my senses and powers of observation are very limited.

My city friends know well enough what I do here during the bee season; it may seem strange work to them, but it is indisputably work; what I do during the slack times is harder for them to figure out: "organizing my ignorance" is perhaps as good a description as any.

One of my projects during this midsummer pause in the beework has been trying to sort out the St.-John's-worts. Some of the species within this family Hypericaceae are maddeningly similar. They are to be found all over the eastern half of the country and in the northwest, too, and most have clusters of bright yellow blossoms that make them look like small sunflowers. Specimens of several of the species grow down by the river, and there I can puzzle over them in between cooling dips; others grow in upland pastures, where farmers consider them a nuisance, because, although stock rarely graze on them, when they do they can cause skin irritation and loss of condition "especially [to] white animals," a sober U.S. Department of Agriculture guide to common weeds tells me. Why *white* animals, I wonder? That sounds irrational, magical. Can the U.S. government be less than reasonable?

St.-John's-wort is a plant to which a good deal of folklore still clings. One of the reasons it is such a nuisance to farmers today is that their predecessors, valuing it for magical reasons, encouraged it to grow and planted it around their houses. In this country it is a naturalized immigrant, not a native, but it crossed the ocean with the reputation of being a prized herbal remedy and a specific against "phantastical spirits."

European peasants hung it in their windows on St. John's (Midsummer) Eve to avert both the evil eye and spells cast by the spirits of darkness. Its Italian name is "devil chaser." When gathered on a Friday, it was said to keep off devils and lightning, but it had to be treated with respect. In some places it was believed that if a person were to step on the plant in the dark of night, a phantom horse would rise from the roots and sweep him up on his back, then gallop away until dawn. The dew that fell on the plant the night before St. John's Day was held to be efficacious in preserving the eyes from disease, and the entire plant was collected, dipped in oil and made into a balm for wounds. As I own no white animals, I am glad that superstition has preserved the plant and helped it spread. I like the St.-John's-worts, which are pretty plants. The shrubby St. Johnswort, *Hypericum spathulatum* in particular, with its compact masses of golden flowers, is handsome enough for a cultivated flower bed.

T. wonders about honey, and says there ought to be something about it in this book. Perhaps a recipe, too.

The nectar of flowers is over eighty percent water, and the sugars in the nectar are complex. To make honey, the bees must evaporate the water and invert the sugars—change them from complex to simple. The bees suck the nectar up through their long tongues and store it in a sac called a honey stomach. When this is full, they fly back to their hives and transfer the nectar to the young house bees, who spread it, drop by drop, throughout the honeycombs in the hives. One bee would have to fly the equivalent of three orbits around the earth in her foraging flights (using one ounce of honey as fuel for each orbit) in order to produce a single pound of honey. In the process of collecting nectar, storing it in their bodies and transferring it to the house bees, the bees add enzymes to the nectar. These break down the complex sugars into simple ones, chiefly dextrose, levulose and sucrose.

The water in the nectar evaporates slowly from the droplets spread out through the hive, but the bees speed up the process by fanning with their wings to create currents of air from the hive entrance at the bottom to the ventilation hole at the top.

When most of the water is removed from the nectar, the bees cap each cell of finished honey with snow-white wax that is secreted in flakes from their wax glands. This finished honey has a very low moisture content, less than nineteen percent, as dry as parched corn, dryer than air. This makes honey hygroscopic: because it can pull moisture from the atmosphere, it must be stored in tightly sealed containers once it is extracted. It is the reason why baked goods made with honey stay moist and do not dry out the way they do when sugar is used.

Dr. Johnathan White, who has worked for the U.S. Department of Agriculture as a honey specialist, gives the following percentage characteristics of an average domestic honey:

Component	*Average*
Moisture	17.2
Levulose	38.2
Dextrose	31.3
Sucrose	1.3
Maltose	7.3
Higher sugars	1.5
Total acid	.57
Ash	.169
Nitrogen	.041

The exact composition of honey would vary from the above chart because plant nectars vary. Different nectars also determine the color of honey, its flavor and the speed with which it crystallizes. For example, bees working buckwheat flowers produce a dark, strong-flavored honey; those working clover make a mild white honey. The blackberry blossoms my bees were working in May are the source of a light-colored, fruity-tasting honey, while the reddish honey with a deep, rich flavor comes from the sumac they were working during the summer. Bees work the best nectar source within a two-mile radius of their hive; in most places, a variety of flowers are in bloom at the same time, so that their honey generally contains a mixture of plant nectars. Honey that is labeled "clover," "orange blossom" or "tupelo" usually contains honey from other flowers as well.

There are other animals who like honey in addition to bees. Ants, cockroaches and wasps get into beehives to eat it when they can.

So do pigs, bears or any other animals with a sweet tooth. Humans are the most skillful at taking the honey from the bees—and they like to spread it on hot biscuits. Honey can replace an equal amount of sugar in many recipes, but other liquid ingredients must be proportionately reduced. When baking with honey, oven temperatures should be lowered by about twenty-five degrees to prevent overbrowning.

The other evening I was reading through a stack of 1920s agricultural magazines I had found in an abandoned cabin up the road. They were the only things left after vandals had stripped the place of its windows and doorknobs, but they made pleasant reading, and I didn't need the doorknobs. In the February 1924 *Better Farming*, a delightful and instructive periodical, I came across the following:

> Busy Bee Helps Motorist—The Rural Engineering Department and office of the extension specialist in agriculture of the New York State College of Agriculture at Cornell University have been conducting some tests with commercial sweets as anti-freeze mixtures for automobile radiators.
>
> Results so far indicate a boiled mixture of equal parts of honey and water to be far ahead of any other combination. This mixture gets slushy at 1.4 degrees Fahrenheit but does not freeze solid.
>
> Low grade honey was found to be just as effective as a high grade. The honey mixture need be put in the radiator only at the beginning of winter, water being added to fill the radiator when the mixture becomes low . . .

This would work, of course, because honey doesn't freeze; nonetheless, I was surprised at such an innovative use, even though I shouldn't have been: people have employed honey for purposes other than eating for centuries—it has been an ingredient in the centers of golf balls, in shaving creams, shampoos, gear lubricants, chewing tobaccos and gum. It was used to embalm the dead in the Egypt of the pharaohs.

"Hooni cleareth all the obstructions of the body, looseneth the belly, purgeth the foulness of the body, and provoketh urine," said Charles Butler, who not only wrote about the keeping of bees in 1623 but was a melliphile as well. "It cutteth up and casteth out phlegmatic matter and thereby sharpens the stomach of them which by reason have little appetite. It purgeth those things which hurt the clearness of the eyes and nourisheth very much; it storeth up and preserveth natural heat and prolongeth old age."

Many claims have been made that honey lengthened life. An ancient Polish king once attributed his one hundred and thirty years to the keeping of bees and the eating of honey. Staunch old king.

Honey ranks among the most astonishing of cure-alls: its antibacterial qualities have made it valued as a dressing for wounds and burns, and there are those who say that it increases the hemoglobin count, prevents anemia, contains an antihemorrhaging factor and aids the body in the absorption of calcium. Others swear it will allay coughs, relieve fevers and inflammatory infections and make an excellent gargle. Honey has been suggested as a cure for radiation sickness, drunkenness and hangovers, while in India it is mixed with beeswax and prescribed for ulcers. Farmers have been known to treat their cows' mastitis with honey. (It should be noted, however, that certain melliphobes have whispered that "sourwood honey . . . is considered to have undeniable griping qualities." They also say that it is the cause of crib death in infants less than six months old.)

Notable mellivores include Caruso, who always downed a tablespoonful of honey before singing; Hindu males, who, according to custom, are fed honey at birth; athletes of all ages; racing pigeons; cows in need of an easier calving—and, of course, the Greek gods, whose original food, legend has it, was the honey of Hymettus.

Unscrupulous vintners have been known to doctor improperly

ripened wine with honey. From ancient times, mead has been brewed from honey, and buckwheat honey was used to make beer. Honey used to be smeared on salted meat to improve its taste. A mixture of milk, honey, salt and butter do honor to guests, and honey has always been considered an appropriate offering to the gods.

Honey has also been used as a spray adherent, a cure for pipe bowls and as a plant growth stimulant. Honey will serve as a bait for houseflies. It is used to line petroleum storage tanks, as well as to preserve transplant tissues. Eggs in cold storage are sometimes kept in honey; in fact, its properties as a packing material are so good that it has been used for shipping plant grafts, seeds and birds' eggs for scientific study, while boars' sperm frozen and then stored in honey is said to possess greater motility when it is thawed.

People who buy my honey use it for many unusual purposes: several insist that a teaspoonful mixed with the local pollen keeps them from sneezing during the hayfever season, and one very old Ozark hill woman takes a tablespoon along with two teaspoonfuls of black pepper every day. She says it keeps her free of arthritis. One young man tells me that for the past six months he has been living on nothing but my honey, distilled water and the juice of pressed clover leaflets. He praises the fine surge of energy that comes from downing a whole cup of honey at a time—he's a regular melli-maniac. I have also been told that the replacement of honey by white sugar in the diet has been responsible for many human disasters, from the defeat of Napoleon's Grande Armée in Russia down to the generally spongy qualities of today's television generation.

Well, they are all my customers, bless them, and they can do anything they want with my honey in the privacy of their homes, but I work hard to help the bees make a fine-*tasting* honey, and it is rather a letdown to find it is being bought for its *moral* qualities.

Here, for T. or anyone else who would like to try it, is a good pie to make with honey:

HONEY APPLE PIE

Pastry for double-crust, 9-inch pie
3/4 cup sugar
1 teaspoon nutmeg
Enough pared and sliced apples to fill a 9-inch pie generously
1 1/2 tablespoons butter, cut into small pieces
1/2 cup liquid honey
1 tablespoon grated orange rind
Confectioners' sugar

Preheat oven to 425° F.

Prepare pastry sufficient for a double-crust, 9-inch pie. Roll out half the dough and line the pie plate.

Combine sugar and nutmeg; pour over apples, lifting and tossing with two forks until well combined. Pile fruit into pie pan, heaping to make a nice fat pie, and dot with butter.

Roll out remaining pastry and cut into 1/2-inch strips. Arrange strips lattice fashion over apples, pressing edges down firmly. Bake 10 minutes, then lower the oven temperature to 350°F. Bake 30–40 minutes more, or until apples are tender and crust is brown. Remove from oven.

Combine honey and orange rind, and pour mixture through openings in lattice; return pie to oven and bake another 5 minutes. Cool to lukewarm and dredge with confectioners' sugar. Serve warm or cold.

Oddly enough, bees are not as sensitive to sweet tastes as we are. Karl von Frisch discovered that they are unable to distinguish between pure water and a three-percent sugar solution, which tastes distinctly sweet to us. There is good reason they evolved as animals unable to detect low concentrations of sugar, however, for such would not be good provision for winter stores, which must be concentrated and compact. The bees' excitement and efficiency in food gathering increases directly in proportion to the sugar concentration of the source. Foraging bees choose the sweetest nectar available, and ignore the others. Honey—one of the most highly concentrated of all sweets—excites them most. If a few bees from different hives discover an open super of honey in the back of a pickup, they will immediately fly to their hives and enthusiastically recruit their sisters; in no time the super will be filled with bees greedily ripping open the honeycomb cells and robbing the honey, even killing one another for the chance to get at it. There will be a veritable bee war over the spoils. And that puts the bees in a bad temper.

During the honey harvest, when open supers are put in the backs of pickups, bees are often touchy and cross anyway, because it is a time of dearth. Honey is harvested at different times in different parts of the country, but whatever the month, it is when the nectar flow is over, which means that there is a full work force of perhaps 60,000 bees staying home in each hive, with little to do. There are often so many bees that they will not fit inside the hives, and they cluster dispiritedly out in front. They are mature bees with full venom sacs, and they are cross, easily provoked to defensive stinging behavior just at the time the beekeeper wants to steal their honey.

The important thing is to take the honey from them as expeditiously as possible and transfer it speedily to an enclosed bee-proof place.

There are a number of ways to take honey from bees and bees from honey. A beekeeper with just a few supers will have no trouble removing all of them from his hives quickly, even if he just sets them to one side and then, frame by frame, shakes the bees from them. But the procedure does stir up the bees, and if he has more than a few supers he will need another method. A bee brush will make the job faster and disturb the bees less. A bee brush is a wooden-handled brush with long, soft, yellow plastic bristles, which he may buy from a bee supply company. Using it, a beekeeper can brush the bees gently away from the combs. Bees do not seem to mind being brushed off nearly as much as being shaken.

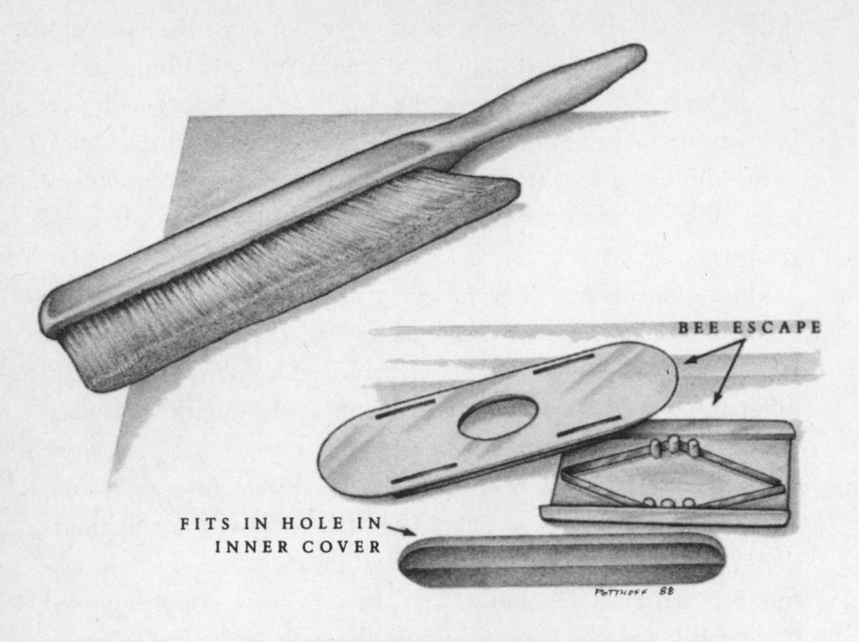

Bee brush and bee escape

However, when a beekeeper uses a bee brush he still must take frames out of the super one by one. If he has a lot of supers, he will prefer to remove the bees from a super all at once.

"Hmpf! Playtoys!" said an old beekeeper the first year I was keeping bees and asked him about bee escapes. Bee escapes are small one-way doors that can be pressed into the hole of an inner cover placed below the top super. The bees can go out of the bee escape but they cannot enter it. When a beekeeper uses bee escapes during the honey harvest, he must put the escape in an inner cover and place that under the uppermost super. In a few days the bees will have left. He can then remove the super and put the inner cover with the bee escape in it under the next super, wait a few days, take the super and continue the process until all of the supers on the hive have been removed. The method is not only slow but requires extra handling of equipment, and for that reason it is unsatisfactory to many beekeepers. I tried it one year on a couple of hives as an experiment, but it was so slow I did not repeat it. Yes. Playtoys.

Bee repellent used on a "fume board" will drive bees from an entire super, but on hot sunshiny days only, manufacturers' claims to the contrary notwithstanding. A fume board is a cloth-lined ridged cover cut to the dimensions of an open super. A beekeeper squirts a small amount of repellent on the cloth lining and places it on top of the super. The fumes are heavier than air, and the bees run from them. The beekeeper takes the cleared super and puts the fume board on the one below it, repeating the process until all the supers have been removed. This is a fast method. but the chemicals

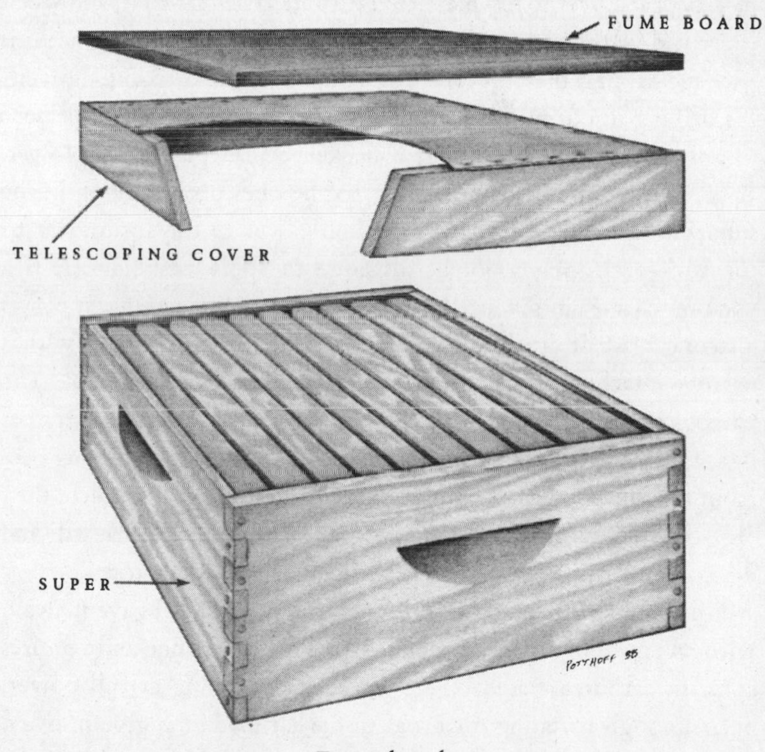

Fume board

do not work well on cold or cloudy days so it is a method that must be reserved for proper weather.

Benzaldehyde, or artificial oil of almonds, and butyric anhydride which is sold under the name of Bee Go, may both be used as repellents. The bees dislike both, and I can see why—even to a human they smell terrible. Benzaldehyde is marginally less offensive to my nose but it is volatile and dangerous enough so that it cannot be shipped through the mails or by UPS, so I do not like carrying it around to my beeyards in my pickup. Butyric anhydride can be shipped, but it smells worse. One of my bee-supply catalogs refers to it as "pungent." I should call it stinky. I once loaned my fume board and butyric anhydride to another beekeeper, who was unable to drive back over to my place to return it, so he mailed it, and when it arrived in town the postmaster phoned: "Don't know what this package has in it that just came for you, but it stinks to high heaven! It's driving us out of the post office. Will you come get it, please?"

For several hundred dollars a beekeeper can buy a bee blower. A bee blower is a machine similar to the ones people living in the suburbs use to blow leaves from their lawns in the autumn or to the back-pack sprayers orchardists use to apply pesticides. It is a vacuum cleaner in reverse. Powered by a gasoline engine, it blows a strong blast of air down a length of flexible hose tipped with a narrow attachment that looks like a vacuum's crevice tool. The narrow tool is used to direct the full force of air between frames in a super, to push the bees downward. It is effective in driving bees from the supers quickly, and does not irritate them overmuch: they regard wind, even strong wind, as one of life's natural hazards and do not rush out to sting the beekeeper creating the storm.

For some years I used a bee blower, but eventually I gave it away, with an enormous sense of relief. It is a fussy machine and requires constant carburetor adjustments to keep it running at full power, but, although irritating, that was not what made me give it away. No, my reason was an aesthetic one: I wearied of the noise of the engine, which is louder than my chain saw, louder than my brush cutter, louder than a lawnmower. It seems wrong, somehow, to make such a terrible racket in a peaceful, beautiful beeyard. It does not put the bees in a bad temper, but it puts me in one. I now use a pair of fume boards and butyric anhydride to harvest my honey. My harvest is in August, when the weather is hot and the chemical works quickly. By using a pair of boards, I can pry up the cleared supers from one hive and hand them to the young man I hire as a helper while the fumes are beginning to act on the second hive. We can usually work through a beeyard with ten or twelve hives in half to three-quarters of an hour. And, after using the blower, I appreciate the quiet in the beeyards enough to put up with the stench of butyric anhydride.

Tony, the young man who has worked for me during past honey harvests, helped me again this year. He's in college now, and I had feared he would have other things to do, but he likes working with the bees and arranged his summer vacation schedule in order to be free for the first few weeks in August, when I needed him. He grew up on a farm, is used to hard work and is comfortable and relaxed around all animals, including bees. Because he has worked for me before, he knows all the routines. He is cooperative, cheerful and calm. He is the best helper I have ever had. When I am running fewer hives, I hope I shall be able to handle the honey harvest on my own, but it's companionable to work with Tony in the beeyards and in the honey house. We drive to the outyards together in my three-quarter-ton truck, which will carry a 5,000-pound load. He stacks the filled supers on the truck, and in the honey house he helps separate the honey from the honeycomb inside them. At first, we worked in the outyards, and after three days the honey house was filled with stacks of supers on pallets. We still had more to bring in, but the honey house was full, and so on the fourth day we began extracting honey.

Even with just a few hives of bees, it is worthwhile for a

beekeeper to make up some pallets the right size for honey supers and use a handtruck for moving them. For me it is absolutely essential. I do not know who invented the handtruck, but whoever he was I hope he led a happy life and was rewarded for his ingenuity. All over the world there are people who have been spared heavy lifting by his invention; I am one of them, and I often look at my three handtrucks and smile.

Tony stacks the supers in the back of the three-quarter-ton truck directly on to pallets piled six high if he is going to move them, five if I am. A stack five high weighs approximately three hundred pounds, and even with a handtruck that is my limit. But Tony is bigger and stronger, and he is usually here, so he generally stacks them six high. Back at the loading dock he uses one of the hand-

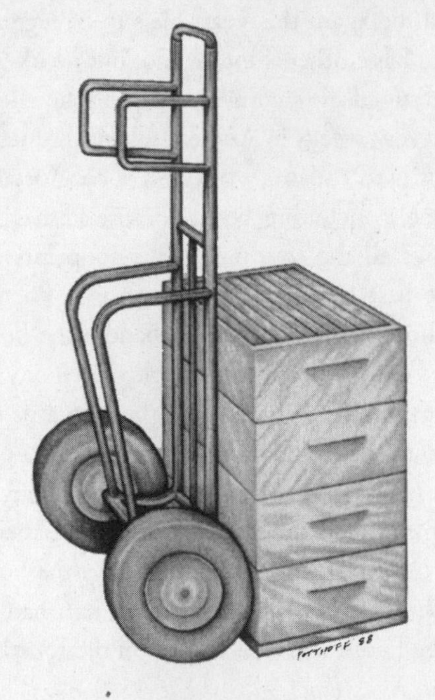

Handtruck

trucks to wheel them off the truck and into the honey house, where they stand until we are ready to start processing honey.

Temperature and humidity are important once honey has been removed from the beehives, and the bees can no longer control them. Honey processors with bigger operations than mine have a hot room, where temperature and humidity are regulated while the filled supers are stacked. I don't worry about temperature, because August is ordinarily so hot that the honey will flow easily from the combs. But I am concerned about humidity, so I do not like to leave the supers any longer than three days in the honey house lest the honey in them picks up moisture from the air.

Whether a beekeeper has three hives or three hundred, as I do, or three thousand, extracting the honey from the combs is much the same job. In all cases, he needs to cut off the wax cappings from the frames of honeycomb so that the honey beneath them is exposed, to take the honey out of the comb and put it in something else. The equipment needed in any sized operation, then, are variations on a hot knife, to slice through the soft wax cappings on the frames of honey; a centrifugal spinner, to separate the honey from the comb; and a container to catch the free honey. Beekeepers with thousands of hives have fast and expensive machinery with which to do this, but a beekeeper with just a small number of hives can, for a few hundred dollars, outfit himself with a knife he can warm up in a basin of hot water, a hand-operated extracting machine and a couple of stainless-steel buckets. My own operation is an in-between size. I use a motorized, steam-heated uncapping knife, a motor-driven extractor that will hurl the honey out of forty-five frames in fifteen minutes and a series of stainless-steel settling and bottling tanks connected by clear plastic tubing.

When we start to process the honey, Tony wheels out the first stack of supers with the handtruck and takes up his position at the uncapping knife. At dawn I had lit the bottled-gas hot plate under the old pressure canner I use to generate steam. The steam passes down a hose into the knife, where it heats the blade so that it can melt the wax. The exhaust from the knife is captured in another hose, which is used to warm a series of copper pipes in a tank into which I dump the cut-off cappings. Melted down into creamy yellow blocks, I will sell this beeswax to a beekeeping-supply company for more money, per pound, than I ever get for the honey. Beeswax has many uses besides making fine candles. It is a base for ointments, creams and cosmetics. It is used for waterproofings and polishes. It is an ingredient in adhesives, crayons, chewing gum, inks, ski and grafting wax. But its chief value is to beekeeping-supply companies, which remelt it and mold it into new frame foundation to sell back to beekeepers.

Once the uncapping knife is hot, Tony switches on its electric motor. This makes the knife move up and down vertically, and when he passes a frame lightly against it, the heat of the blade melts and neatly saws through the delicate wax cappings without tearing the honeycomb cells. Then he flips the frame over and uncaps the other side. The uncapped frames, oozing with honey, are placed temporarily in the tank that also catches the wax cappings. Honey drains from the frames and the cappings into another tank, which feeds into the rest of the system. When he has forty-five frames uncapped I load them into the extractor—a big circular tank fitted out with a slightly smaller heavy screen basket that is slotted to hold the honeycomb frames. When I turn on its motor, the basket starts spinning and the honey inside the frames is thrown out against the walls of the extractor, where it runs down and out through a series of tubes and baffled settling tanks into one of the three 1,200-pound bottling tanks. The baffles hold back most of the bits of loose wax that break off from the frames, but just before the honey runs into the bottling tanks, it is strained through a double layer of nylon mesh to keep it clean and free of any remaining wax particles.

When one 1,200-pound tank is full, I switch the tube leading into it to one of the others and empty the tank. When I am extracting honey, I process all of it without heat and empty the bottling tanks into sixty-pound plastic containers with tightly fitting lids to keep out moisture. Throughout the year, before I make sales trips I will warm the honey enough to break down its crystals. Unheated honey tastes better than heated, but unheated honey crystallizes, and American consumers prefer honey in liquid, not crystalline, form, so I must heat and bottle it before I can sell it.

But now we are simply taking honey from the combs as rapidly as possible, to preserve its flavor and quality, and putting it into sixty-pound storage containers. We have set ourselves the goal of running twelve extractor loads through the system each day. We can do that before the worst of the August-afternoon heat makes the job too uncomfortable. Some supers are fuller than others, but normally it takes us about four or five days to extract all the honey we have harvested in three. When we are finished, we head back out to the beeyards and fill up the honey house once again, repeating the whole process until every super is emptied of honey.

Harvesting and extracting honey is hot, heavy, repetitious labor. Because bees are lively, complicated and unpredictable animals, and therefore interesting, and because honey is just honey, I find this part of the operation more like work than the rest of it. I am grateful to have such a hardworking and good-humored partner as Tony to help me.

Over the past five years Tony has made his spending money by cutting cordwood, and so as we drove around to the beeyards we talked about wood and woodcutting. I told him I had cut most of my firewood for winter, but that there was one tree out in back of my home beeyard I was going to have to call in one of the neighbors to fell. It is a big tree, and the sixteen-inch bar on my chain saw is too short to go through it in a single cut. I've never dropped a tree so big, and I was afraid of making a cut on each side. On the last day Tony worked for me in the honey house he brought along his own bigger chain saw.

"When we're done in here," he said, "I'm gonna go out and drop that big tree for you."

And he did, too. The best helper I've ever had.

When Tony leaves in the afternoon, I like to go back into the cabin to shower away the layer of stickiness and weariness that has accumulated from a day of working in the honey house. As the shower pours over me, I think what a fine plan it is that honey is soluble in water. My feet ache from standing on the concrete floor of the honey house, so after my shower I like to prop them up, drink a cup of coffee, enjoy the comfortable tiredness that comes after a day of hard physical labor and contemplate my good fortune at actually being able to make a living by associating with a bunch of bugs.

Later on, at sunset, I have more work to do. The honeycombs still have a small amount of honey left in them after they are removed from the extractor. It is too little to try to remove mechanically, but too much to leave in the combs when they are stored for the winter, because it would attract cockroaches and ants. So each morning at dawn, before the bees start flying, I have loaded into my old half-ton truck the supers we processed the previous day and driven them out toward the woodlot. During the day the bees have removed every last trace of honey from them. On the first day I took out a load, the bees, of course, did not know it was there. But within a quarter of an hour they were sure to have discovered it and recruited their sister workers. During the day, thousands of bees worked feverishly on the cleanup. Having learned the position of such a splendid source of food, they were waiting for me the next morning when I drove up, circling impatiently in the air above the spot where the truck had been. By the fourth day, they knew the time I'd be there—5:30 A.M.—and they kept their appointment with me and the pickup all during the rest of the harvest. After the last day that supers have been cleaned up and stored, I know they will still wait expectantly in the same spot for several mornings, hovering in the air in the dawn light at the spot where the truck has been parked in days past. At sunset—as the bees start to return to their hives, and before the nocturnal wax moths start flying—I have been walking out to the truck each day to drive the load of clean supers back to my storage shed, where I unload and fumigate them.

Wax moths are a problem for beekeepers who store empty comb-filled equipment in all but the most northerly parts of this country. Without bees to keep them in check, wax moths who can find ways into the supers will lay their eggs, which will hatch into hungry larvae. If a beekeeper stores comb without fumigating it to kill them he will find nothing but the ruined skeletons of comb, ropy webbing where the wax-moth larvae have worked, and frass and cocoons nested into the wooden frame parts when he opens the supers up in the spring. So I fumigate as I stack. I use para-dichlorobenzene crystals, one of several fumigants available. The vapor from the fumigant is heavier than air and will kill larvae below it. I stack three supers on a pallet, tear off a corner of newspaper, put a tablespoonful of para-dichlorobenzene on it, place it in the center of the super on top of the frames and then cover the top with a layer of newspaper to seal in the fumes. I repeat the process above the newspaper layer until I have reached the roof of the shed and have a stack twelve supers high. I place a cover above the final newspaper sheet so that mice cannot get into the stack and make nests.

The para-dichlorobenzene fumes kill live wax-moth larvae, but they do not affect the eggs adult wax moths may have laid, so in ten days I will have to repeat the process to kill any new larvae that may have hatched out in the interim. I leave no supers uncovered, and I seal up any cracks or holes I find with duct tape so wax moths cannot squeeze in and lay new eggs. After the second fumigation, the stacks of supers should be safe from wax moths if they are not disturbed in any way. The para-dichlorobenzene crystals break down rapidly in warm weather, and after a few days the fumes dissipate, so by springtime, when the supers are ready to use again, there will be no residue left in them.

Yesterday evening, I brought in the last load of supers from this year's honey harvest. I was in a hurry, because I had other work to do in the evening, and so I went out for the supers a little too early. It was still light, and I knew there would be many bees left on the supers. The day had been hotter than usual, and after my shower I had slipped the most minimal of sundresses over my nakedness and called it getting dressed. But before I walked out to get the supers, I pulled on a pair of heavy socks and workboots to walk through the pasture back to the truck. I often see copperheads on that open ground. Copperheads are the shyest, least aggressive snakes I know. Their dearest wish is to slither away from a human being, but if I were to step directly on one, he would strike to defend himself. A copperhead's bite, although not fatal, is poisonous enough to make me sick, and so I wear boots when I walk through the pasture. In my sundress and Mammy Yokum footwear I must look the real hillbilly.

Not only honeybees, but wasps and bumblebees as well remained on the supers in the truck when I got to it. Some of them blew off as I drove back to the storage shed. I knew that the ones remaining were so single-minded and intent on their work that they wouldn't bother me while I unloaded the truck, so I climbed up on it and began shaking each super free of bees as I unloaded. The bees swirled around, hardly noticing me as they worked and as I worked. When the bees were shaken from the supers they fell to the truck bed, and some of them began flying up underneath my sundress. I could feel them crawling across my skin. The bees were in a good mood, to be sure, but it was asking too much of them not to sting when they discovered they were trapped under the dress. I didn't want to be stung, so I peeled off the sundress. Once they were freed from the confining cloth, they flew away peacefully and I continued to unload the truck inside a cloud of bees, naked as a jaybird except for my big workboots.

The late-afternoon light had changed to dusk when I was done. I pulled on my sundress and drove the truck under the pine trees, where I had left it. Another honey harvest was ended.

I started to walk back to the cabin, but the cries of nighthawks made me stop and look up. Nighthawks are related to whippoorwills and chuck-will's-widows and are not hawks at all. With their long, pointed wings, the underside of which are banded with white, they are easy to recognize, and are often seen around city parking lots, where they feed on the insects attracted by overhead lights. I sometimes see them here in the evening, swooping after the first of the night-flying insects but not often in such numbers as they were last evening. Their calls—*peeee-eent, pe-ee-eeent*—filled the air. They were gathered not to feed on bees—although they were probably doing some of that too—but to migrate. Late August still feels like summer here in the Ozarks, but it is the time of year the nighthawks are moving on to their South American wintering grounds.

Our human calendars take little notice of such dates, but the nighthawk migration tells of shortening days and a season's end. The honey harvest marks it, also. The end of the bees' year is the beginning of the bees' year. It is mine, too.

Afterword

February 14, 1988. Valentine's Day. I am a beekeeper but I am also a writer, and some years ago I sat down at the typewriter to experiment with words, to try to tease out of the amorphous, chaotic and wordless part of myself the reason why I was staying on this hilltop in the Ozarks after my first husband, with whom I had started a beekeeping business, and I had divorced. I wasn't sure why I was here, and because I am a rational creature and like to know what's going on, I wanted to process what I was doing through my brain cells, to put it into words and see if I could arrive at an understanding of it. I had always written for an audience, but this time I was writing for myself, and what I put on paper over the next couple of years was unlike anything I had written before. I traced the natural history of my hilltop from one springtime to the next, discovering by the second spring that I was in a new place and understanding the value of where I was. It was a little like minimalist music. Each day resembled the one preceding it; the steps were barely noticeable, but the end was different from the beginning. I had made a record of changes I had thought too subtle for me to have noticed until I started writing.

I had not been writing for publication, and I was astonished when a friend told me I was writing a book. My agent said I was writing a book, too, and in due course my editor at Random House said I was writing a book. After it was published it even looked like a book. And I was still astonished, and remain so to this day. Once put out into the world, that supposed book took on a life of its own; it became a part of the process of change in my life.

I was in town one day after it was published, picking up my mail. The letter on top of the pile had a return address in Washington, D.C., under the name of an old friend from college. Seeing his name opened a door of memory so easily and so happily that it was hard to believe I had closed it as firmly as I thought I had more than thirty years ago. It was the oddest feeling. "Of course," I thought to myself. "Of course."

In his letter my friend said he had seen a review of my book in *The New York Times* the previous Sunday. He had not known my last name, the name of my former husband, but he had still recognized me after all these years from the picture accompanying the review. He had bought the book and sat up all night reading it. He briefly sketched what had been happening in his life in the years since we had gone our different ways. Would I answer his letter? I did. But before he had had a chance to receive my reply he had telephoned. There has seldom been a day since that we have not talked on the telephone or been together. He came visiting. We canoed. We took trips. We swam. I stayed with him in Washington for longer and longer stretches of time. One thing led to another, and yesterday we were married.

His work and his interests keep him, for the most part, in Washington. Mine take me to two places. I want to be with him in the city, but my Ozark hilltop and its wild things and wild places pull me all the time. This probably sounds complicated and it is, although in quite practical terms it will be less complicated by the time I can reduce my beekeeping down to the hundred hives I want; they will produce just enough honey to take care of my established accounts in St. Louis, Boston, New York and Washington. The frequent, time-consuming sales trips to other places will be over. Selling honey has always been the least appealing part of beekeeping to me, anyway. It is a very complex life, but, after all, I've never wanted a simple one; keeping the various aspects of it that change brings about into some sort of harmony is the liveliest and most interesting kind of life I know.

I delight in moving back and forth between two ways of living, one reflecting off the other, clarifying both in the process. I like knowing how to use the subway as well as how to bring proper function back to my 1954 pickup when it becomes balky out in a remote beeyard. I like seeing both experimental theater on Fourteenth Street and deer grazing winter grasses at twilight on my Ozark hillside. I like playing dress-up and going to an embassy party to watch people who get their names in the newspapers parry with one another. And I like pulling on a baggy bee suit, forgetting myself and getting as close to the bees' lives as they will let me, remembering in the process that there is more to life than the merely human. I like company and learning to say "we" again. I like solitude. I like loving a man and I like being loved by him.

Glossary

Balling. Bees' method of killing a queen by surrounding her in a compact mass.

Bee brush. A wooden-handled brush with long soft plastic bristles that can be used to gently remove bees from places their keeper does not want them.

Bee dance. The patterned motions by which bees communicate tactilely.

Bee escape. A small one-way door that fits into the hole in an inner cover, and which may be used to allow bees to leave one part of their hive but bars them from returning to it.

Bee space. A distance of a quarter to half an inch that bees will not fill with comb.

Beeyard. A place in which bees are kept in hives.

Bottom board. The rimmed platform on which hive bodies stand.

Brood. Eggs, larvae and developing immature bees.

Cleansing flights. Flights bees make from their hive to defecate.

Colony. A group of related bees that live together as a unit.

Drawn comb. Honeycomb that has been fully worked by bees; the foundation cells have been built up into deep hexagonal wax cups to receive either nectar and pollen or the queen's eggs.

Drone. A male bee.

Entrance reducer. A block of wood or other material that closes off most of a beehive entrance.

Frame. The supporting structure for honeycomb within a hive.

Frame grip. A grasping tool used for removing single frames from a beehive.

Fume board. A rimmed board cut to the outside dimensions of an open beehive top. It is covered with fabric, and will absorb liquid bee repellent. Placed fabric side down, the fumes from the repellent will drive the bees downward into the lower reaches of their hive.

Hive. Technically this word refers to the wooden boxes and their parts in which bees live, but it is often used loosely by beekeepers as well as by the public as a synonym for "colony."

Hive body. The basic unit of a beehive, which contains the comb on which bees live and work.

Hive staple. A large tinned wire staple used for holding together beehive parts while they are being moved.

Hive tool. A short stout prying implement for loosening beehive parts.

Ill tempered. The state of bees who are quick to sting. May be caused by temporary stress, such as poor weather or queenlessness, or by genetic predisposition, such as that found in wild (German) bees or Africanized bees.

Inner cover. Board on top of hive bodies but under the telescoping or outer cover.

Laying worker. A worker bee who lays nonfertile eggs in a queenless colony.

Nuc. Rhymes with duke. A small beehive containing three or four frames used as a starter for a standard ten-frame hive.

Nuptial flight. The flight from her hive made by a virgin queen to mate with drones.

Outyard. A place where bees in their hives are kept, away from the beekeeper's principal place of operation.

Package bees. Bees with a queen gathered by a bee breeder and shipped in a screenwire cage.

Propolis. A gummy substance manufactured by bees, principally from plant resins, used by the bees to seal up the inside of their hive. The word is derived from the Greek, and means before (*i.e.,* in front of) the city (the bees' place of living).

Queen. The fertile female bee who lays eggs from which all the bees in a colony grow.

Queen cell. A pendant, peanut-shaped cell, larger than those in which worker bees or drones develop, in which the queen pupates.

Queen excluders. Rigid welded wires placed in wooden frames cut to the outside dimensions of the hive body and super tops. The wires are close enough together to keep the queen from going through but far enough apart to allow passage to the smaller worker bees. Drones cannot pass through them, either.

Queenright. The condition in which a colony of bees finds itself when it has a vigorous, well-mated queen who is capable of laying eggs.

Robbing. The stealing of honey or nectar from a hive by worker bees from other colonies. Robber bees, as my daughter-in-law, Liddy, phrases it, are "females gone wrong." They often have a sleek, greasy appearance.

Smoker. A hand-held metal firebox attached to a bellows that is used to generate the smoke which quiets bees as their keeper works with them.

Supers. The cases for honeycomb frames that are shallower than hive bodies and sit on top of them.

Supersedure. A process by which the bees raise a new queen to replace one who is no longer adequate to their needs, and who is termed by beekeepers to be a "failing queen."

Supersedure cell. A queen cell, often midway down a frame of brood.

Swarm. A group of bees with a queen that has split away from a parent colony to fly off and establish itself in a new place.

Swarm cell. A queen cell, often at the bottom of a frame, that is a possible indicator that bees are raising new queens to accompany them when the colony swarms. There are usually several of them at a time.

Tanging. Making a loud noise, usually by beating metal against metal, in order to supposedly bring down a swarm of bees flying overhead.

Telescoping cover. The usual overtopping cover to a beehive employed by all but migratory beekeepers, who use cleated covers instead.

Worker bees. The majority of bees in a colony. They are females with atrophied sexual characteristics who gather nectar and pollen, manufacture propolis, raise young bees, defend the hive, build comb and tend the queen.

Index